S0-AZB-069

# Praise for *Almost Human*

As well as any book I have read, *Almost Human* captures both the science of the animal and its being. The baboons are individuals, not merely statistical entities; each has its own temperament and idiosyncracies . . . Dr. Strum combines art and science uncommonly well to describe the mystery of another creature. Books such as hers create an awareness that animals too have unique and complex societies as worthy of study and preservation as our own.

—from the foreword by George B. Schaller

Full of wonderful insights into the behavior of baboons, scientists, and other animals. In lucid and entertaining prose, Shirley Strum takes us into the life of a baboon troop, displacing simplistic myths about dominance and aggressiveness in favor of a subtle portrait of how sophisticated social animals work out their relationships. *Almost Human* succeeds in being both moving and deeply instructive—as well as being an extraordinarily good read.

—Philip Kitcher, University of California

A courageous book which personalizes one primatologist's attempt to better understand the minds, the social lives, and the "nature" of the animals she studies—work conducted with a tremendous dedication to the well-being of the animals and to her own vision of the science of primatology despite the intermittent lack of acceptance by a sometimes inflexible academic community.

—Linda Fedigan, University of Alberta

An inspirating scientific and personal exploration of a previously misunderstood primate group.

—*Kirkus Reviews*

Shirley Strum's evidence that baboons evolved alternatives to aggression, that is, a friendly, peaceful society, is quite remarkable . . . *Almost Human* ranks with Jane Goodall's *In the Shadow of Man*.

—*Washington Post*

This is one of those rare books where you learn as much about the making of science as about its results. Primatologists and primatology taken together make a fascinating story.

—Bruno Latour, Center for Sociology of Innovation, Paris

This is an honest, in some ways an outrageous book because it treats together things that happen together that convention keeps apart: science and politics (in the widest sense) and the life of the scientist. I wish I had written it myself.

—Dr. Thelma Rowell, *Natural History* magazine

"Gives more insight into the life of both baboons and primatologists than anything I've ever encountered."

—Dr. John Fleagle, SUNY Stony Brook

# ALMOST HUMAN

Drawings by Deborah Ross

Shirley C. Strum

# ALMOST HUMAN

## A Journey into the World

## of Baboons

W. W. Norton
New York • London

*To Jonah, Laura and the baboons*

Copyright © 1987 by Shirley C. Strum
Introduction copyright © 1987 by George B. Schaller.
Illustrations copyright © 1987 by Deborah Ross.
All rights reserved.

First published as a Norton paperback 1990 by arrangement with Random House, Inc.

Library of Congress Cataloging-in-Publication Data
Strum, Shirley C.
Almost human.
Bibliography: p.
Includes index.
1. Olive baboon—Behavior. 2. Mammals—Behavior.
3. Mammals—Kenya—Behavior I. Title.
QL737P93S79   1987      599.8′2      86-29712

ISBN 0-393-30708-5

Printed in the United States of America

W. W. Norton & Company, Inc., 500 Fifth Avenue, New York, NY 10110
W. W. Norton & Company Ltd, 10 Coptic Street, London WC1A 1PU

3 4 5 6 7 8 9 0

*The Bushman . . . indicated that to
tell [the story] properly one should use the
language of the baboons; however, he added
respectfully, "I must speak in my own
language, because I feel that the speech of
baboons is not easy."*

LAURENS VAN DER POST, *The Heart of the Hunter*

*I confess freely to you, I could never look
long upon a monkey without very
mortifying reflections.*

WILLIAM CONGREVE *in a letter to Jean Baptiste Denis*
(1695)

# Foreword

During the past quarter century, hundreds of scientists, most of them graduate students, have penetrated remote corners of the world in search of an interesting species to study. Among them was Shirley C. Strum, who in 1972 set out for Kenya to observe olive baboons. She wanted to learn how they meet the challenges of savannah life in order to gain insights into the way early human society might have adapted to a similar environment. In the normal course of events she would have spent a year amassing a pile of unadorned facts and impersonal descriptions that would then be tabulated, graphed and finally recorded in a thesis, using all the proper jargon. Science would have been served.

But to interpret another culture, one which cannot speak and which leaves no artifacts, requires more than skill with vital statistics and glib scientific notions. In any study, the delight of watching animals should be primary, the details secondary. As Poincaré wrote:

empathy for baboons. As many studies have shown, aggression is adaptive, a way of dealing with a competitive situation. However, such behavior may be disruptive, and societies, especially stable ones in which individuals know one another, have evolved social alternatives to aggression. When studying mountain gorillas in the late 1950s, I was impressed with the peaceful nature of their society. Other studies—on wolves, African hunting dogs and several monkey species, to name a few—all stressed low levels of aggression. Yet baboons, as a result of brief early projects, were viewed as having societies in which males had a rigid hierarchy and achieved their ends by force—behavior which was then ascribed to human society.

Nevertheless, Shirley's interpretations created a furor among some in the narrow circle of primatologists concerned with baboons. Had she presented her ideas to, let us say, a great ape or carnivore clique, she would have received a sympathetic hearing. The inertia of the human mind, Arthur Koestler rightly noted, is most clearly demonstrated among those professionals who have a vested interest in tradition. Shirley had begun her study within a certain conceptual framework, but she still looked at the baboons with fresh eyes, and when her observations collided with her preconceived ideas, she willingly accepted a new vision.

If a scientist takes too much vocal pride in objectivity, beware. Observing is subjective: the animal described is only an illusion created out of a personal perspective, based on which questions are raised, which facts written down, which information ignored. Another biologist asking different questions will create a different animal. The conspicuous, easily described behavior is turned into statistics; the difficult but no less real behavior tends to be ignored or considered irrelevant. To describe another being takes not merely reason and fact, but also empathy and intuition. Not all biologists have the understanding and courage to become heretics, willing and able to describe animals as they truly perceive them. *Almost Human* does this superbly.

I wonder how some of the outstanding early ethologists like Konrad Lorenz and Niko Tinbergen, who so obviously were enchanted with animals, view the recent scientific trends in their field. The mental abilities of animals have often been denied, rare but significant behavior has frequently been discarded in favor of columns of numbers; it has even been claimed that animals lack awareness—that without language they cannot think. I once watched a lioness kill a gazelle during a communal pride hunt, then sit nonchalantly by the carcass hidden in

The Scientist does not study nature because it is useful to do so. He studies it because he takes pleasure in it; and he takes pleasure in it because it is beautiful. If nature were not beautiful, it would not be worth knowing and life would not be worth living. . . . I mean the intimate beauty which comes from the harmonious order of its parts and which a pure intelligence can grasp.

Good research needs both a longing to understand and emotional involvement. This Shirley brought to the task. The result is one of the few field studies that provides a satisfying insight into a species.

Ideally, a researcher should be close to the study animals, but not interact with them or affect their routine. With time and patience, Shirley was accepted by one group, the Pumphouse Gang. Baboon males are only transient members of groups; newcomers are integrated slowly, as over many months they move from the periphery to the core of the group and there establish friendships. Unwittingly, Shirley used the same correct technique, and was accepted. She accompanied the animals from dawn to dusk, so that they became "friends, in a very unusual sense." The gentle anecdotes and elegant interpretations of family and group life of the individuals she came to know represent the most fascinating chapters of *Almost Human*.

Shirley found that the baboons had become "a passion, almost an obsession," and from her extended study she gained not just knowledge but understanding; she ceased being a scientific technician focused on a narrow aspect of baboon behavior and became a humanist who considered the philosophy, ethics and science of her involvement with the animals—aspects that give this book its unusual depths.

Until recently, an intensive study of a year or two was considered long enough to provide the relevant data of a species. However, every animal has a personal history, its behavior influenced by previous contacts with group members, relatives and neighbors. Since a human observer cannot interview the subjects, he or she can only watch patiently as one generation comes and another passes away. This almost means ceasing to live for years on your own terms and instead accepting the animal's reality, something few biologists have the tenacity to do. Shirley's important discoveries are the result of over a decade of dedication.

She found that baboons live in a peaceful society in which not aggression but friendship achieved the desired results. Her insights will have a far-reaching influence on how primate societies are perceived; her descriptions are immensely appealing, creating new respect and the grass at her feet until the other lions wandered away and she could claim her prize without sharing. Shirley discovered "extraordinary intelligence, planning and insight" in her baboons. These animals were most definitely aware of the consequences of their planned acts. A loving dog-owner can tell you more about animal awareness than some laboratory behaviorists. I find it humbling to contemplate the truly amazing number of relevant pieces of social information a baboon can store in its brain and quickly retrieve so that it can respond with finesse to any one of thousands of potential social interactions in a large group. Scientists with a penchant for teaching animals that which they can do least well might feel uncomfortably lacking in IQ if they tested the mental prowess of baboons with an array of social questions rather than with bits of plastic of different color and shape. Like several other recent works, *Almost Human* recognizes and accepts mental abilities in animals, an important reversal of a scientific trend.

Baboon society is not, of course, a template for human society, early or otherwise. Olive baboons, Hamadryas baboons and patas monkeys—all three of them savannah dwellers—have different social systems; males leave the group in baboon society, females in gorilla society. Such studies as *Almost Human,* however, also help to define the spectrum of options open to early humans and the principles that might have governed their societies. This is more than an intriguing scientific exercise, for it emphasizes that humankind is not bound by its heritage into using force, that it can choose to live in peace, harmony, cooperation and friendship. A contemplation of baboons can help humankind correct a skewed vision of itself.

As well as any book I have read, *Almost Human* captures both the science of the animal and its being. The baboons are individuals, not mere statistical entities; each has its own temperament and idiosyncrasies, each has its own desires and goals. The baboons inspired Shirley to become passionately involved in their existence, and her lively pen describes the treasure of secrets she uncovered. Despite their value, scientific papers cannot express the fundamental charm, the fleeting social entanglements, the perishable moments of a baboon's life; they cannot deepen our love and understanding of another species, they cannot establish a heartfelt unity with creatures that were once part of our past.

Albert Einstein wrote: "The most beautiful thing we can experience is the mysterious. It is the source of all true art and science." In *Almost Human,* Shirley C. Strum combines art and science uncommonly well to describe the mystery of another creature. Books such as hers create an awareness that animals too have unique and complex societies as worthy of study and preservation as our own. With more and more species crowding the brink of extinction, they also need chroniclers, not merely to write obituaries but to fight for their survival with insight, eloquence and a sense of outrage. One aim of studying animals is, after all, to prolong their lives.

GEORGE B. SCHALLER

# Acknowledgments

My baboon research has gone on for so long and benefited from the help of so many people and agencies that it is impossible to thank everyone concerned. But let me make a start with apologies to those unmentioned others whose involvement is also greatly appreciated.

My work in Kenya was made possible by the kind permission of the Government of Kenya and through the sponsorship of the Office of the President, the National Museums, the Institute of Primate Research–Nairobi, the Department of Wildlife Conservation and Management and the University of Nairobi. Over the years, the money to support my research has been provided by: The National Science Foundation–U.S.; the University of California, San Diego; The L.S.B. Leakey Foundation; The New York Zoological Society; The World Wildlife Fund–U.S., The Fyssen Foundation–Paris; The H. F. Guggenheim Foundation and The National Geographic Society.

A variety of individuals have been involved in some aspect of the baboon research at the doctoral or postdoctoral level or as research assistants, and have contributed to the project records from 1976 to the present. While the views presented in this book are totally my own, I would like to thank the following individuals for their participation and, in some cases, also for their friendship and help over the years: Fred and Denise Bercovitch, Bob Byles, Neil Chalmers, Kris Cunningham, Tag Demment, Deb and Bronco Forthman-Quick, Hugh and Perry Gilmore, Bob Harding, Linda Hennessy, Julie Johnson, Joe Kalla, Susan Lingle, Lynda Muckenfuss, Nancy Nicolson, Debbie Manzolillo Nightengale, Ronald and Bette Noe, Mary O'Bryan, Tom Olivier, Gloria Petersen, Mike Rose, Mark Saunders, Linda Scott, Barb Smuts, Debbie Snelson, Jeff Swift and Matt Wilson.

Jim Else, director of the Institute of Primate Research, pitched in to help the baboons at more than one critical point and I owe him special thanks. My husband, Jonah Western, was often the man behind (and sometimes in front) of the woman. Without his insights and help I couldn't have weathered the multitude of crises.

The Cole family, the Dansie family, the Morjaria family and Stefano Cheli extended more hospitality than I can ever hope to repay. Carleto Ancillato and John Jessel made the translocation possible. Kuki Gallmann, Iain Douglas-Hamilton, and Ker and Downey Safaris helped at crucial junctions.

Tim Ransom and Bob Campbell documented the life of the Pumphouse Gang and hangers-on photographically, providing me with both wonderful company and outstanding stills and cine film.

In recent years the majority of the daily hard work of observing the baboons has been shouldered by a group of Kenyan research assistants. I want particularly to thank Josiah Musau, deputy director of the Uaso Ngiro Baboon Project, and Hudson Oyaro, chief research officer, whose outstanding work and cheerful support have made it possible for me to continue with the research even during periods when I couldn't physically be with the baboons. Thomas Kingwa, Lawrence Kinyangui, David Ooga and previously Simon Ntabo and Francis Malile complete the Kenyan research team.

Throughout the past two decades, Sherwood Washburn's rich intellectual framework was my guide. I hope that he will be happy with the fruits of his labors. In more ways than they know, Thelma Rowell, David Hamburg, Hans Kummer and Bruno Latour have had a major impact on my ideas about science, baboons and other primates.

Finally, this book is a reality only because my husband insisted,

Laura Nathanson helped and encouraged, Candida Donadio had faith, Jean-Isabel McNutt worked tirelessly and my colleagues in the Department of Anthropology, University of California, San Diego and the university administration made it possible for me to live my life both as professor in California and baboon researcher in Kenya.

And thank you to my parents and to the baboons who, each in their own way, had patience and understood my dilemma.

# Author's Note

This book is fact, not fiction. I have made certain alterations in order to keep the cast of characters small—it includes over four hundred baboons—and to help the reader grasp the most important points about the animals. These changes in no way alter any truths or impressions about baboon behavior. The names of human actors have sometimes been changed to protect both the innocent and the guilty.

S. C. S.
*Nairobi, 1986*

ALMOST HUMAN

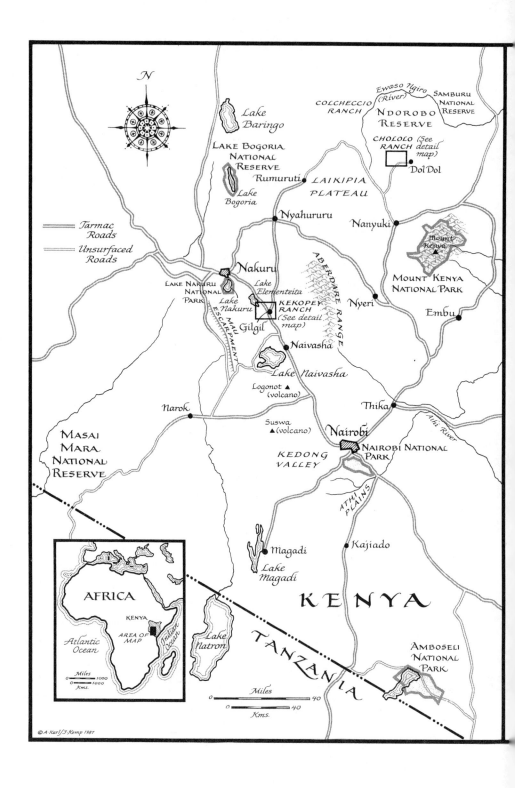

© A·Karl/J·Kemp 1987

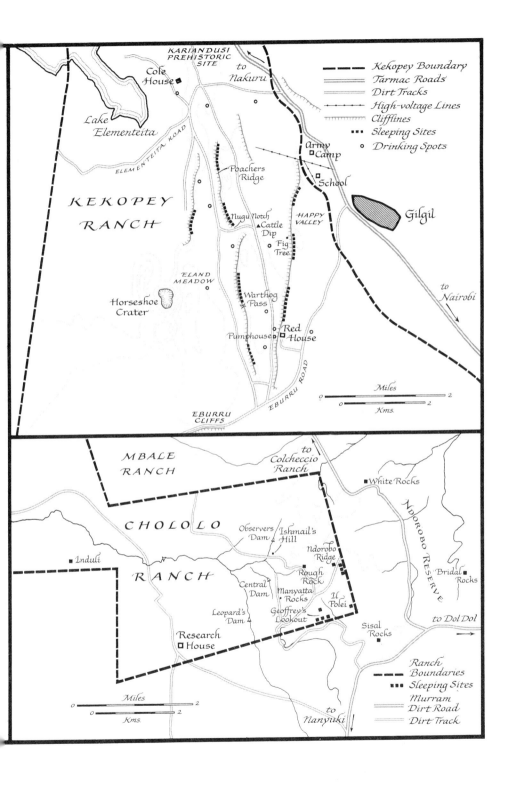

# I. Starting Out

I was hunched uncomfortably into the corner of a VW van, bumping along a rutted road chaotic with ancient cars and donkey carts. A truck passed us, listing heavily and barely missing our sides. With irony, my driver translated the Swahili slogan painted on it: "No hurry in Africa."

Well, I thought drowsily, if that was true, I'd have a lot to overcome. It was September 10, 1972, and my money would last only until January 1974. Here came another slogan-carrying vehicle, a particularly ramshackle taxi-bus—a *matatu*—with something I could translate for myself: "*Hatari Safari.*" "Dangerous Journey." So far I was simply exhausted, not frightened. With the San Diego–Nairobi flight behind me, I was finding this leg of the trip, from urban Nairobi to rural Kenya, the most draining. We had quickly whisked through the quaintly provincial city of Nairobi and into surrounding farmland. A patchwork of tiny fields covered a steeply rolling terrain, with deep river gorges suddenly appearing here and there. The van lurched with agonizing slowness along an uneven ribbon of asphalt. Clusters of toiling people surrounded us. Women, their black faces shining with sweat, were bent forward as they walked with huge stacks of wood on their shoulders, the loads held taut by a thick plaited rope bound around their foreheads. Even small girls moved along with burdens I could never have managed. My neck muscles ached in sympathy. Melodic, incomprehensible voices and animal noises filled the hot, still air. Quickly the scene changed again. Lushness gave way to open aridness; fields invaded a thick bush not unlike the California chapparal. I nestled next to my suitcase, trying to find a comfortable position, gave up and closed my eyes, finally dozing off.

Miles later, I was jerked awake. My driver announced that it was time for some fresh air. Once I was out of the van, the Great Rift Valley of Kenya stretched before me; I could see for at least forty miles in both directions. There were no buildings, no people, just endless vistas. In the distance were the extinct craters of Longonot and Suswa, rising from the floor of the Rift. The scene was daunting. Not even the Pacific had prepared me for the dimensions of the African savannah. This was the Africa I had imagined, not the steaming jungles of common fantasy nor the clog of humanity we had just experienced, but vast grasslands teeming with wildlife. This was the landscape of human evolution. Some say the savannah's special opportunities—its wildlife—attracted our earliest ancestors from the forest. Others suggest that humans were forced out of our original forest home by more successful competitors: monkeys and apes. But it doesn't matter whether we were bright opportunists or desperate fugitives. The feat was monumental. How had the earliest humans managed? Vulnerable and primitive, they had only rudimentary tools, no language and a brain not much bigger than a chimpanzee's. We know a lot about their bodies and their anatomy, but the real key to their survival was their behavior. What was it like? How had they overcome the challenges of their new environment?

This was more than an academic question. We, today, are their descendants, products of an experiment that began three or four million years ago. We have inherited both their talents and their shortcomings. If, in this modern world, we hope to realize our human potential, maximizing strengths and circumventing limitations, we have to know what is in our evolutionary heritage.

It was this quest that had brought me to Africa. In recent years, science has begun to tackle the problem of human origins. Paleoanthropologists have studied and interpreted our predecessor's fossils. Anthropologists, zoologists and psychologists have examined the behavior of our closest living relatives, the nonhuman primates.

Already some scientists felt they had discovered answers about our past. Humans began with an aggressive society that was tight and cohesive. It had to be; there was safety in numbers. But protection went beyond density. Aggressive, powerful males provided a formidable defense against large savannah predators. There was bound to be aggression and competition within the group among the many males as well. Male jostling created a dominance hierarchy in which every male had and knew his place. This hierarchy prevented constant aggressive contests. Size, brute force and dominance status created male leaders who determined where the group went and what it did.

Females played their part. They were the mothers, bearing children and caring for them. But their attention was always focused on the males, the critical core of the group. Females didn't need male political skills since they had little to do with group protection or leadership.

These ideas had impressed more than the scientists. Robert Ardrey and others told the lay public that modern humans were not far removed from our primeval days. A killer ape still lurked inside each of us. Men and women were naturally different in abilities. Even if we wanted a society with greater sexual equality, we might not be able to overcome millions of years of biology.

What if they were right? Worse still, what if they were wrong and we believed them? As a student in the new field of primate behavior, I was in Kenya to study olive baboons, *Papio anubis*. Like the early humans, baboons have met the challenges of savannah life. Few other primates have. Watching baboons might help us understand the problems early humans may have faced and the solutions they found. Baboons are not relics of our human past, yet comparing the options open to two primates on the savannah is more productive than making comparisons between humans and elephant, wildebeeste and lion.

If these weren't reasons enough to study baboons, there was another critical one. The model of an aggressive human society controlled by powerful males was an *extrapolation* from the first studies of baboon behavior. Could we really make the link so simply between baboons and humans? Apes are more closely related to us in their biology and anatomy than are monkeys. Which species should we take as our template, chimpanzees or baboons?

I agreed with those who had selected baboons. The human adventure began with our shift to the savannah. It was almost unique in primate history. Almost. Baboons took that fateful step and survived. What could chimpanzees or gorillas or orangutans tell us about what it must have meant?

Although I sided with baboons, I wondered whether baboon society was really governed by aggression, dominance and males. There were grounds for doubt. The pioneering first studies had their shortcomings. They identified only the few adult males, gave them names and watched them carefully as individuals. Females and youngsters were shortchanged, lumped together by age and gender. Perhaps these individuals were more interesting, complex and important than they seemed. I expected so, based on what had been learned about the variety and complexity of behavior in other closely related monkeys and apes in the years since those first baboon studies.

What are baboons really like? How useful are they in our quest to understand ourselves? This was what I hoped to discover.

As we walked back to the van, I was told we were not far from Kekopey, my destination. Kekopey was a 45,000-acre cattle ranch, the home of "my" troop of baboons. These wild animals had already become accustomed to humans. Three years earlier, Bob Harding, the first graduate student to watch them, had christened them the Pumphouse Gang, after the Tom Wolfe book. The name stuck. Like their human namesakes, they spent a lot of time around a pumphouse situated near one of their sleeping sites.

I hung on for dear life as we bumped along toward Kekopey. I was tired and excited. I was also beginning to worry. Now that I was here, armed with my burning intellectual questions, could I really accomplish my task? The realization that I was an unlikely person for such a job was finally sinking in. I was properly trained, with lots of ideas. But what about the rest? Was there anything in my past that indicated I could manage a study of wild baboons?

I was a city girl, born and bred. To make matters worse, I was not athletic or physically active. I'd been camping only once in my life and the first night had been a disaster; I let my imagination run away with fantasies of dangerous beasts about to attack. I'd been a rather isolated only child, always looking for a connection with something bigger. History provided a tangible link to the past for me. I had felt it most strongly when, at the age of eleven, I was taken to the Acropolis. I had experimented with religion too, first a Jewish cultural heritage going back five thousand years and then other religions: Taoism and Buddhism. But it was not until I reached Berkeley as an undergraduate in September 1965 that I finally found my niche. As I lived through the Free Speech movement and the Vietnam War protests, I found myself confronting, over and over again, questions about human nature, about what was innate and impossible to modify and what was flexible and worth changing.

For a while I toyed with abnormal psychology, then sociology. But it was my first cultural anthropology class that finally convinced me: here was the right approach. The course looked at human behavior from a cross-cultural perspective. One class led to another: I decided to major in anthropology.

Then the other shoe dropped. I sat with a thousand other students, mesmerized, as Sherwood Washburn traced the human inheritance back further and further, to the earliest prosimians, those least progressive primates of sixty million years ago. As Washburn held the fossil of a tiny prosimian in his hand for the class to see, I marveled that this minute creature had experienced the world, when it was alive, more the way I did than my own dear cat, Crazy. Like me, it perceived its surroundings in depth and in color. Its delicate primate hands already had finger pads and nails, and its brain had changed and grown to control them. I felt linked not only to a few thousand years of art or culture, but to millions of generations, to something bigger than I had ever imagined existed.

Sherwood Washburn was a soft-spoken, small man. His salt-and-pepper hair went perfectly with the modest glasses and Ivy League dress that reflected his East Coast upbringing and Harvard education. All other stereotypes stopped there. When he spoke about the essentials of evolutionary principles or about human evolution, he cast a spell. He was no showman, doing tricks to entertain and amuse; he simply knew his subject—he created much of the field during his lifetime—and enjoyed it with a fervor and intensity that was contagious. Until I attended his introductory class on human evolution, I knew little and cared less about the subject. Suddenly it seemed the most important way to understand human behavior. His version of the evolutionary perspective offered me exactly the answers I was seeking—a rich approach to the topic of why we are the way we are.

It would probably embarrass Washburn to know the extent to which he became my guru, but I was not alone. A long string of undergraduates and graduates, before and after me, reacted in the same way. For several decades, Washburn contributed more students to the profession than anyone else, and I joined them, going to every seminar, every lecture that was available.

But I had a problem. Although those fossilized remains told the basic story, I found the bones rather boring. It was the *behavior* they implied that interested me. I couldn't see myself dissecting rotting cadavers and becoming an expert on bone shafts, or sitting in the blazing sun painstakingly excavating small bits of fossilized bone with a toothbrush.

My salvation came when I discovered how useful the behavior of living primates can be in reconstructing the evolution of human behavior. It was a new field, and only a few universities offered graduate training programs in it. I entered Berkeley's graduate school in September 1969, so that I could stay close to Washburn and his remarkable insights.

———————

Again the van came to a sudden stop. I lurched forward, my reverie interrupted. It was my first close look at wild animals, and they nearly stepped over us. Three magnificent giraffes were sauntering across the road. Gracefully tall, dancelike in the rhythm of their stride, with soft hairy mouths, doelike eyes and fuzzy, stumpy horns, they were an unbelievable configuration of legs and necks. Our driver sat impatiently racing the motor until the last millimeter of giraffe was past, then rushed on. When you've seen one giraffe, his acceleration said, you've seen them all. I hoped never to become that blasé.

And yet what did I really know about the kind of life I was about to enter? Certainly it had not been the romance of the wild that lured me to Africa. In fact, I was ignorant of the wild. Most of what I knew about animals or nature came from books. My two childhood pets, a short-lived goldfish and a parakeet who could sometimes be persuaded to say the five words he knew, and later, Crazy, the cat, did little to initiate me. Lacking firsthand experience, without great physical skills or a strong desire for adventure, I could only hope that my intellectual passions would help me through the rest of the challenges.

We stopped again. This time it was engine trouble. While too many people tampered with the motor, I had a another look at the grasslands. The savannah was as impressive close up as it had seemed from the edge of the escarpment. Although there were lots of animals, even with my binoculars I had difficulty really seeing them. I'd spot an interesting silhouette and then it would evaporate in the intense heat haze. Others would appear, but it was hard to tell whether they were a mirage or real. I located a herd of zebras. They, too, vibrated in and out of focus, the heat playing a dizzying game with their stripes. There were rust-colored antelopes of several sizes; I couldn't tell whether they were male and female of the same species or two different species. Some other big animal was also mixed in with the zebra and antelope. This was certainly different from watching monkeys in large cages at the Berkeley behavior station, or animals in a zoo or safari park. Merely seeing and identifying the animals was a challenge to my inexperienced eyes.

I wondered if there were any patas monkeys living here. I tried to remember the maps of their distribution. Besides humans and baboons, patas are the only modern-day primates inhabiting the open veld. Vervets, the small green monkeys with black faces, hands and feet, live in trees along savannah rivers but never really venture far into the open.

I winced slightly as I thought of the patas. They are also known as hussar monkeys, because of their military-style mustaches and the patterning of their strawberry-blond-and-white hair. I had originally wanted to study these beautiful animals. Patas intrigued me. They live in small groups with only one adult male in each group, and although the males are more than twice the size of the females, they do not dominate group life. Groups are led by the females, who also act as the policers. The male's role is to be on the lookout for possible predators. Taking as high a vantage point as he can find in a tree or bush, he alerts the group of danger, then puts on a diversionary display to attract the predator's attention and runs as fast as he can in the direction opposite to that of the group. Meanwhile the rest of the troop remain silent and motionless in the tall grass, hoping to escape detection. This would be a foolhardy defense if it weren't for a special patas adaptation: male patas are remarkable runners, probably the fastest primates in the world.

Patas survived life on the savannah in a manner completely opposed to the described life-style of baboons. They seemed an anomaly. Why did savannah living produce one set of behaviors in baboons and another in patas? What would it have taken to make our human ancestors more pataslike and less baboonlike? These were questions I proposed to answer in the patas study I wanted to do. At that time, only one brief study of the patas, by Ronald Hall, existed, and I felt optimistic that my proposal would be well received.

Finally, Washburn called me to his office. I could tell by the expression on his face what was about to follow; still, I was shocked. He rejected my idea totally. Patas had been done! There was nothing of great note left to learn, he proclaimed. I agreed that Hall's work was superb, but he had watched patas for only six months. We now considered one thousand hours of observation the minimum standard. Hall had raised important questions, but we still lacked many of those answers. I thought I had produced a convincing case.

By the time I had enough courage to see Washburn again, to ask for his suggestions, I was feeling very low indeed.

BABOONS! I couldn't believe my ears. What did he mean, baboons? They were the *most* studied species of primates. They were the basis of the existing model of early hominids, the one I was dubious about. I couldn't think what else there was to learn about baboons.

Given the situation, there was only one avenue left to me. I decided to salvage the ideas I would have explored in the patas study and adapt them to baboons.

I never learned why Washburn accepted baboons and not patas, but I doubted that even he realized what a productive avenue my baboon study would prove to be.

At the time, though, I wasn't thrilled. I knew about baboons already, from the classes I had taken. The baboons I had seen in films and at the zoo had not impressed me. They had none of the appealing attractiveness of the patas. In fact, in medieval literature, baboons represented the repulsive—the evil spirits, not the good ones. Even today to call someone a baboon is a grave insult. In Kenya, and elsewhere in Africa, baboons are not thought of as wildlife or as valued and protected animals. Legally they arè vermin. But I had to admit that this attitude was actually a tribute to their success as a species. Until recently, there were many more baboons than people in Africa, some said. No matter the exact figures, baboons were numerous. I could admire their success, but wondered whether I could ever feel more positive toward them.

Since I was interested in intellectual issues, not cuddly animals, it shouldn't matter, I told myself. Often in my life, even as a child, I'd used intellectual pleasures to transform an unpleasant, difficult or threatening situation. Maybe it would be the same with studying baboons.

Excitement returned as I wrote up my proposal. Whatever imperfections there were in our understanding of baboons, these animals were of critical importance to interpretations of human evolution. I wanted to "test" the baboon model again, to see whether baboons were as they had been described.

How was I going to do this? I had two tools. The first was to apply new observation techniques to studying the species. These were more rigorous and systematic than the methods Washburn and the others had thought necessary in the era of the first baboon studies. I would also identify all the individuals in the group, not just the adult males, as the early baboon watchers had done. By identifying all the individuals, watching a representative sample of them, giving females equal time with males and using the new observation techniques, I hoped to eliminate any selective biases that might unconsciously have been incorporated into the previous picture of baboon society. Perhaps I would learn something more about how males and females functioned in the group.

Washburn felt that observation alone was not enough. If I wanted to find out how important the males were to the group and what roles females played, I'd have to experiment. At the end of my observational study, I had decided to trap all the males and temporarily hold them captive. That way I could see what the remaining, untrapped females would do. Who would lead, police, protect the troop? Where would the group go, and how would they fare with the neighboring baboons?

As the VW bounced closer and closer to Kekopey, I only hoped that The National Science Foundation, the agency that was funding my project, knew what they were doing with their money. I had with me all the necessary study tools: binoculars, tape recorder, cameras, books about primates and papers describing the previous baboon studies. But would that be enough? In retrospect, I see that it was a strange combination of intellectual fervor, naïveté and a complete ignorance of what it meant to live in Africa that made the whole process of getting to Kekopey merely another turn of the academic grindstone.

A week later I lay back in the clear hot bathwater and contemplated Africa. The sarcophagal bathtub was almost six feet long; I had to point my toes to keep my head from slipping under. The bathtub was one of the many surprises that first week at the Red House, or "Kiserigwa" as it was called in Swahili. The Red House had been built by the owners of Kekopey for a short-lived ranch foreman, then abandoned, and was donated to Bob Harding and his family when they came to study the baboons in 1970. It had needed work: doorknobs and windowpanes were missing, and the only furnishings were the layers of bat droppings.

It was a funny house, full of incongruities, and it had seen better days. Thick stone walls and lots of windows made it cool and airy. The inside whitewash was dingy from a constant smoky pall, a combination of fumes from the kerosene refrigerator and the wood-burning stove. It had once been wired for electricity and telephone, but both had disappeared during the last twenty years. The kitchen boasted a splendid lime-green Dover stove, and the badly marked floors showed the remnants of still-beautiful parquet. Candles and hurricane lamps provided light at night. The stove provided hot water. When it burned for several hours, there was enough hot water for one luxurious bath or for several quick and miserly ones. Chopped wood provided fuel for cooking and heating water; laundry was done in the bathtub and ironed with a charcoal iron on a rickety ironing board.

Compared to city life in California, this was roughing it. But compared to a tent pitched beneath an acacia tree in the middle of nowhere, it was sheer luxury. The odd U-shaped layout of the house forced everyone to walk through the main bedroom to get to the bathroom and the other bedroom. Right now it did seem a little crowded. My host, Matt Williams, and his wife shared the big bedroom. I had come to replace Matt, who himself had replaced Bob Harding. We were all students of Washburn. I shared the small second bedroom with three others. Lynda was also a Berkeley graduate student, who was studying black-and-white Colobus monkeys and hadn't yet found a place to live. Tim Ransom was a friend of mine from Berkeley. Having just completed his doctorate in psychology, studying baboons at the Gombe Stream in Tanzania, where Jane Goodall studied chimpanzees, he had decided to become a wildlife photographer and was shooting pictures on assignment from the *National Geographic*. Squeezed into a third bed was Matt's young daughter. With barely enough room to maneuver in our dormitory-style room, I was grateful for my privacy today; the others had gone to Nakuru, about forty-five minutes away, for supplies.

An Abaluya house-servant-cum-cook made the baboon work possible. Without Joab, much of each day would have been sacrificed to simple household chores. Joab was elderly, wise and kind. He had already ministered to me during my first bout of dysentery. He had tapped gently on my door, and entered holding a pitcher of water. In mime he urged me to drink, realizing more than I did that dehydration was causing the worst of my symptoms. He was right.

The garish orange exterior of the Red House seemed out of place.

It sat on a hill, hidden from below by a dense thicket of tall leleshwa bushes. A magnificent view extended in all directions. On one side was the Mau escarpment of the Rift Valley, with volcanic foothills guarding the approach. Eburru, one of those peaks, steamed on cold days. On another side were steplike scarps that ran the length of Kekopey. Beyond the grassland plateaus I could just glimpse Lake Elementeita. The shallow alkaline water resembled an iridescent mirror edged with a broad pink ribbon. Thousands of pink flamingos, attracted by the abundant algae, lived on the lake. A series of small hills completed the panorama, rising from the faults above the lake. This was Gili Gili—the source of the rivers and Kekopey's water. Water traveled from a hot spring through fifteen miles of underground pipes to a pumping station and then up the cliffline to the Red House and beyond.

The view from the bathtub, through high-level windows, was to become one of my favorites. I could lie in the water and stare at the immense sky, which dominated the landscape in all seasons. This was September, the end of the winter months which are cold, dry and overcast. Today the clouds were large billows of white, partially obscuring the intense blue sky. Even the filtered light made me squint in my bath.

I thought about Matt. I had known him only casually in graduate school. He seemed as unlikely a field worker as I was, being a raconteur and a gourmet cook; his interest was primate communication.

Tim and I depended on Matt for our introduction to the Pumphouse Gang. It had taken us this whole week to find them. Only then did I really understand Matt's assurance that the troop, having been observed and followed by two graduate students since 1970, was now accustomed to being observed.

"You want to get *out?*" he had asked incredulously. "Out? Of the *van?*" It was the white VW van the monkeys accepted, not people. According to Matt, danger lurked everywhere. There were poisonous snakes, vicious warthogs who would rather gore you with their tusks than look at you; crazed male buffalo harboring resentment along with stray bullets, ready to take their revenge on any human; there were even large predators. Matt was firm. We must not leave the van.

––––––––––

At first I thought it was real concern for our safety that had made us so cautious as we drove around Kekopey looking for the baboons. I didn't mind much; I was distracted by the beauty of the place and by the animals. I had gradually begun to see them more readily—herds of zebra, Thomson's gazelles, impala and eland. To begin with I could spot only the largest animals and had trouble telling different species apart. During the last day or two of this week I had begun to spot the smaller, more secretive creatures: foot-high dik-dik and slightly larger rufous steinbok; bat-eared foxes and blackbacked jackals; reebok and klipspringer—both wondrously camouflaged among the gray granite cliffs. There were also warthogs, hyrax, mongooses and another kind of reebok among the more common animals. I saw very few cows; if it weren't for the water troughs and wire fences, this could be wild Africa, not a commercial ranch.

The sky was alive with birds, not the little brown ones that had bored me in California but brilliantly colored ones, the smallest and the largest I had ever seen. Green-white-orange bee-eaters, orange-black-white hoopoes and lilac-turquoise-white rufous rollers were the first I recognized. Even the starlings were beautiful. Magnificent birds of prey soared on thermals and nested on rocky cliff ledges. There were ugly birds as well: giant ground hornbills, their appalling red wattles resting on sinister black chests, and marabou storks, whose bald heads, heavy bills and pendulous neck pouches were set off by the beautiful neck and shoulder feathers that were used to make fancy boas for 1920s flappers.

There were baboons, too—far in the distance. Bob Harding had selected Kekopey after surveying sites in Kenya and Tanzania. Louis Leakey, who had worked for many years at the nearby prehistoric site of Kariandusi and knew the place well, had directed Bob to Kekopey. Kekopey had advantages for Bob. It was conveniently located not far from the two-laned road that ran from Nairobi to Uganda, right through the nearby town of Gilgil. Gilgil had thrived under British rule and was dominated by two army camps and full of a variety of shops. Now it had fewer amenities, although the Kenya army occupied the two camps. Gilgil was really a village. It had four streets lined with rows of small shops—*dukas*—all selling the same things: sugar, canned goods, milk, tea, coffee and a few other essentials. A local bakery sometimes produced bread, and the twice-weekly native market sold fresh vegetables and fruit. Two cities were close: Nakuru was twenty miles away, and Nairobi was seventy miles away. By 1972, Gilgil had shrunk to a few hundred people. Large ranches and smaller farms stretched in several directions. South, around Kekopey, there were few people; wildlife flourished. Elsewhere, the land was being carved into smaller parcels. More people meant less wildlife.

A dirt road from Gilgil ran past the track to the Red House. Aside from cattle herders, Kekopey had few people. Even the owners, the Coles, were not always around.

The Coles had been among the original white settlers of the Kenya highlands and figured prominently in the history of the area. The previous generation, Berkeley and Galbraith, are mentioned in Isak Dinesen's *Out of Africa*. The Coles' house was eight miles from the Red House, at the northern end of Kekopey. This elegant rural retreat replaced the original small bungalow as the main farmhouse. The house was filled with antiques, some certainly from the original consignment to Kenya at the turn of the century.

Arthur Cole had an enchanting way of mumbling his words, a habit that made him appear slightly and charmingly absentminded. He was hard of hearing, a problem he shared with many of the Kenya whites of his generation, none of whom had used earmuffs when firing guns. He refused to wear a hearing aid. If he couldn't hear you, it was your problem.

Tobina Cole was tall and thin, wore Levi's and boots and was immensely capable. I was told that her mother had had a formidable reputation for independence, and that she and a friend once walked completely around Lake Turkana—then Lake Rudolf—a huge lake in bleak and arid northern Kenya. They survived by driving a herd of goats in front of them, eating the meat and making shoes from the hides. She was rumored to have been part of a gang of rowdies who shot out the lights in the rough-and-ready frontier town of Nairobi.

The Coles ran Kekopey by dividing responsibilities. Arthur masterminded the ranching operation, planning the water scheme, opening up grazing land and fencing large tracts with high-tension wire. Tobina managed the cattle. She crossed local and imported breeds. The Kekopey cattle were both beautiful and resistant to many native diseases. Jomo Kenyatta, the first president of the Republic of Kenya, considered Kekopey a showplace, and sent foreign visitors there to see the potential of his country.

I was both impressed and fascinated. In some ways the Coles were as foreign to me as Kekopey's wildlife or the local African culture. Visiting them at teatime was a special treat. We sat at a magnificent twelve-foot table, drinking tea Tobina poured from a Georgian silver teapot, enjoying homemade scones, breads, jams and cakes. Immaculate servants, in white uniforms with scarlet cummerbunds and fezzes, attended us discreetly. The Coles and their guests wore jeans or shorts and boots—often covered with dust, mud and smelling of the farm-yard—but at least hands and faces were always freshly washed.

The Coles loved wildlife, and were proud of the large number of wild animals on Kekopey. Only the large predators, the cats and hyenas, were seen as a threat and removed whenever they caused trouble. Arthur and Tobina had agreed to allow researchers on Kekopey, including the baboon watchers, which made them unusual among land-owners. But I was uncertain about their attitude toward the baboons themselves. Their statements ranged from sheer disinterest to active hostility. Shooting baboons had been a favorite pastime before the baboon studies began. It was good sport, because the monkeys were smart and would try to outwit the hunters. Even now, when the baboons tangled with their dogs, the Coles felt it necessary to shoot them.

I turned to Matt, expecting him to defend the animals, if only from the standpoint of his own research. He said nothing. I felt timid; my stomach was still queasy and I was not up to the defense. Besides, the Coles probably had a point. By all accounts baboons were hardly charming. Yet I felt distinctly uncomfortable about shooting fellow primates just for sport.

———————

I stirred the fast-cooling bathwater, and worried at the problem. Today we had finally tracked down the Pumphouse Gang. Of course I didn't know one baboon from another, but even I could tell the difference in this troop's tolerance. All the other baboons had fled at the mere sight of the van. As the three of us began to watch, it was as if I ceased to exist. Matt directed all of his comments to Tim, although I was the one about to take over the baboon work. Finally I asked Matt a direct question: Couldn't we move the van a bit closer? The troop was forty yards away and from that distance they looked like dozens of brown bumps spread out over the grassland.

Oh, we could get closer without the troop running off, he'd said, but who knows what the animals might do. To the van? I wondered.

I resigned myself to the distance. While Matt told Tim anecdotes of African dangers narrowly avoided, I sat and watched baboons with my binoculars. I had the eerie feeling of being a Peeping Tom, invading the baboons' privacy and spying on their most intimate behaviors without their permission. The Pumphouse Gang moved from one activity to the next in an orderly way. Daily life seemed to follow a definite routine: sleep, then socializing, followed by feeding, resting, more feeding and more resting. Social activities were interposed, particularly during the midday rest period and just before the day ended at the sleeping cliffs.

Each type of activity was itself composed of many parts, and I began to see just how complicated each one really was. Eating took up the most time. I had expected to be bored—how absorbing is a cow grazing in a field?—but the baboons were fascinating eaters. Now I saw what Washburn had meant about the versatility of the primate hand. The baboons' nimble fingers were into everything, plucking grass blades, digging up roots, selecting the tiniest herb peeking out of the ground cover. Flowers, buds, shoots, berries, seeds and pods all fell victim to these voracious monkeys. They seemed to be eating everything from the ground up. Only later would I discover how selective they really were, how good at choosing the most nutritious part each plant had to offer in each season. Their hands had both power and precision; the opposability of the thumb and fingers was as essential here on the savannah as it had been millions of years ago when tiny ancestral primates climbed into the forest canopy. Some foods had to be excavated from rock-hard ground; others had to be peeled or seeded. The monkeys used their teeth to help, usually ending up with a litter of debris scattered on the ground and hanging from their long facial hairs.

Their feeding postures varied. Sitting in one place was easiest, but only if there was lots of food within arm's reach. Often the baboons shuffled along on their bottoms from one patch of food to another. Grass heads were more easily harvested from a standing position. Walking from one spot to another was seldom a waste of time, since there was usually something to eat along the way.

Some variations on these basic themes were comical. One medium-sized youngster seemed too lazy to sit up. He lay on his stomach, chin resting on the earth, plucking the grass from in front of his eyes. Unfortunately when he tried to put the food into his mouth, he had to move his whole head up and down, since his jaw rested on solid ground. Many of the baboons had a Groucho Marx crouch, freeing both hands to feed. Double-fisted feeding was common. I didn't know whether this was because of the appeal of the particular food, the sheer bulk that was needed or fear that a more dominant animal would grab the place.

The monkeys didn't appear to share food with one another, not even a mother with her baby. Taking a closer look, I saw little that could be shared: a blade of grass, a small piece of fruit, a flower, even a root just wasn't big enough.

Along with eating came resting. Resting postures were more idiosyncratic than feeding positions. Chin on chest, knees pulled up and hands folded calmly in the lap was a popular position. Sometimes a monkey sat back and leaned against a rock or tree, almost reclining. Big males occasionally found a rock in the shade and lay down on their backs. They looked particularly vulnerable in this position, with chin, neck, belly and genitals exposed. Mothers rested with their babies cradled in their laps, both arms gently wrapped around them for support. Resting often occurred in clusters; several baboons sat with sides or backs touching, or making tentative but neighborly contact with hands, tails or toes.

While many rested, others socialized. Youngsters took the opportunity, when the troop wasn't moving, to play. Grooming was also common. Even my inexperienced eye could see in fact what I had been taught in theory: grooming was intense and had several functions. The first was hygiene. The groomer removed insects, grass burrs, dirt, scabs. Grooming helped wounds to stay clean and open so that they healed more rapidly, and kept the thick, coarse hair from becoming an uncomfortable matted mess. Baboons will groom themselves, but when they can, they get someone else to do it. Far from being a disgusting habit, as it has been viewed by many humans from medieval times until the present, grooming is essential to good health when you live in the wild.

But grooming is also a major social activity, an excuse for individuals to be close to one another, a way to establish or reinforce positive feelings about others. A look at any grooming pair illustrated why. The animal being groomed was completely relaxed, often nearly asleep, luxuriating in the monkey equivalent of a good massage. If attention needed to be paid to some particular spot, it was thrust in the groomer's face. Often an extended arm or leg or a tilted head revealed the next spot that needed grooming. Grooming was usually reciprocal, but how this reciprocity worked wasn't yet obvious to me. It seemed that nearly everyone had someone to groom, and in turn had someone to groom him or her.

I began to realize that I was missing a lot by being inside the van. It was impossible to hear all the sounds, to catch all the subtle gestures or to know whether smells played a role in baboon interactions. It was impossible to follow any individual if he moved far through the troop or went out of sight behind a bush, a ridge or up the cliffs.

What did Matt think about the difficulty of watching baboons from the van? Was he confident that he could really study communication this way? I wanted to talk to him about other things as well. After a year with the troop, did he understand the complexities of what went on? Who decided where the troop went or what it would eat? How did the grooming pairs match up? Why didn't the baboons share their food?

Reluctantly, I emerged from my bath; I decided then that Matt was not the person to ask. I wondered if he simply didn't like baboons. Perhaps he shared the Coles' feelings and wanted to keep his distance, not from fear, but because, somehow, baboons make us all feel uncomfortable. Intellectually we recognize that we are linked to these animals, through their behavior and through the evolution of our biology. While we are fascinated and challenged by them, we are also embarrassed and threatened, as much today as in Victorian times, when Darwin confronted the world with his new set of "facts."

I kept my theory in mind the next few days, as we watched Pumphouse from the white VW. While Matt talked about aggression and male dominance, I watched. Based on his comments, I expected to see a series of confrontations, male against male—not necessarily outright fights, but at least some serious bluffing. The troop remained peaceful. Matt pointed out the six big males. He was particularly taken with Big Sam, "the brute" as he called him. Matt indicated that Rad was on his way toward another male, Sumner. Now we'll see some action, he proclaimed. But we didn't.

As I watched, I thought about the task ahead of me. How would I learn to tell the animals apart at this distance? I had begun to recognize a few. Surprisingly, the adult males were difficult for me to distinguish from one another, even though there weren't many of them. The animals I recognized were females; their mannerisms, body build and markings seemed quite distinct. I also knew a few baboon facts that would help me with identifications. At least I could sort the troop into ages and sexes. Baboon ages are a combination of chronological age and developmental indications. Infants, born black, turned a peppery color at around seven months. By a year, their coloring was completely adult—a brown-gray-olive mixture. Female juveniles ranged from two to about four and half years—or until the time they first come into heat or estrus. In contrast, a male is a juvenile from two until he is about six years old, when he reaches the size of an adult female. A period of adolescence follows; animals are sometimes called subadults at this point. Female adolescence begins at puberty, at the commencement of the first sexual cycle, and continues until the birth of the first infant, at which point the female, now roughly six years old, becomes an adult. Male adolescence begins at around the age of six and continues until growth is completed, usually around the age of ten. During this period, male canine teeth develop, testicles descend and enlarge and the mantle of shoulder hair reaches its full size.

Pumphouse was currently a troop of sixty animals: six adult males, seventeen adult females and thirty-seven immature baboons. There were other baboon troops on Kekopey, but Matt didn't seem to know either how many or how big they were.

My developing skills led me to have even more doubts about Matt. Several times I was sure he was wrong in his identifications. That was not Beth but Harriet, not Marcia but Frieda. To begin with I thought *I* was wrong. It was Dieter, the infant male that Matt pointed out to me, who helped me gain confidence in myself. Where, I wondered, was Dieter's penis? I knew that infant penises were not difficult to see; a bunch of black infants running around all had what looked like a pink fifth leg. Finally I spoke privately with Tim, who, after a few minutes with the binoculars, verified my suspicions. Tim and I changed Dieter's name to Dierdre, but never said anything to Matt.

In retrospect, I see that Matt did me a a great service. Left alone, I concentrated on the baboons and began to form my own independent impressions. Given what I saw, or didn't see, I was even more convinced that I had to walk among the baboons in order to do a good study; Matt's performance confirmed this. His timidity made my rudimentary courage seem more substantial. All in all, comparing myself to Matt, I gained confidence that I could manage, that I *could* study wild baboons.

The night before Matt left, we stayed in the van as the animals ascended the steep cliff face to their sleeping ledges. I wondered how they managed not to fall off during their long sleep. They all seemed to prefer the smallest ledges on the steepest slopes. Singly and in clusters, they settled down for the night. The light faded, and the troop gradually disappeared, blending perfectly into the rocks. I lowered my binoculars. It was a special, peaceful moment.

Matt started the engine and heaved a happy sigh. His enthusiastic racing of the motor and joyful humming clearly conveyed his feelings—only one more night of Africa.

# 2. Two Newcomers

Two newcomers joined Pumphouse at about the same time. Both sought acceptance and both were kept at a distance to begin with. One newcomer had a distinct advantage over the other: he was a baboon.

I didn't notice Ray—as he came to be called—at first. I was too busy; Matt Williams had left, I was alone for the first time and I was about to test his most cherished theory. I was going to get out of the van and see what would happen. I coasted the VW to a stop at the usual distance from the troop, my heart pounding. I opened the door and slithered to the ground, sitting on the van-shadowed grass. Everyone looked up. I froze in place and averted my eyes: to a baboon, a direct gaze signals a threat. Sixty pairs of eyes focused on me. I waited. The air was clear and sweet, punctuated by the noises of birds and insects.

The sleeping cliffs lay ahead of me; it was early morning, and the animals had just descended and were scattered in the meadow below.

I ventured a look, and when my furtive glance appeared to cause no alarm, I began to observe in earnest until several nearby animals objected. I looked away. The baboons and I played this coy game for about an hour, by the end of which I could watch them without getting any reaction. I raised my binoculars, careful not to catch the glint of the sun. So far, so good. Then I stood up. This was a mistake. The baboons scattered in all directions.

Chastened, and not wanting to press my luck, I climbed back into the van to begin the daily census, as Matt had instructed. I still needed to concentrate intensely just to tell the baboons apart. I began with the males. The one closest to me was Radcliff, or Rad as we called him, an exceedingly elegant male whose long muzzle ended in a Cyrano-like tip. Next I picked out Carl, then Sumner, who was probably the oldest of the Pumphouse males. The others were too distant to identify clearly, but their size made them stand out from the rest, so I just began counting them. Four, five, six, seven. Seven! I must have made a mistake. There were only six males in Pumphouse.

I maneuvered the van closer and counted again, identifying each baboon by name: Carl, Rad, Sumner, Big Sam, Arthur, Little Sam—and X. I was not mistaken; there *was* a stranger with Pumphouse. Now that I was really looking, it was obvious that something unusual was going on. The strange male sat very quietly and stiffly at the edge of the group, trying, like I was, to appear as unobtrusive as possible. All eyes were on him. No one was feeding or playing or grooming. Juveniles and adult females formed a semicircle around the new male. Now and then, some of the youngsters would rush toward him, then lose their nerve and dash back again. I inched closer, and examined the stranger. Ray was striking. I found all the adult males fairly impressive, their fifty-pound bodies enhanced by a heavy mantle of shoulder hair. Ray's was gold-tipped and, backlit by the sun, it shone like a halo. His elegant black hands, quietly folded, might have been etched. He seemed young, because his dark muzzle was mostly smooth, only a few hairline scars revealing past conflicts. His brows fell somewhere between a V and a straight ridge, giving him a bright and intelligent expression. He looked in top form, robust and powerful.

I settled down for a long day's watch. My task—before I discovered Ray's presence—had been to work on baboon identification. Just for the record, I now re-identified all six Pumphouse males, double-checking each one against Matt's descriptions and my own inadequate sketches. Some of the males were easier to spot than others—Big Sam,

*a new member*

for instance, had a crooked smile, apparently caused by some kind of facial injury. This always made him seem menacing. Then it was time to start on the adult females. A few of them were easy: Peggy, for instance, had a quite recognizable tail, and the dark V on her forehead, as well as some of her mannerisms, were unique within the troop. Naomi was also an easy female to identify, not because she had any special characteristics, but because she always hung about on the periphery of the troop. The baboons did not exclude her; her loner status seemed to be entirely voluntary.

I was to discover that distinguishing the rest of the females from one another took intense concentration and a game of comparison and elimination that would end up consuming weeks. Adults, who ranged in age from about six to thirty-four years for females and from ten years into their twenties for males, comprised less than half the troop. The rest—the babies and juveniles—were all very similar in shape and size. Fortunately, however, the youngest babies were easy to identify, because they stayed close to their mothers. Although each one had a pink, wrinkled face, large pink ears and fuzzy, charcoal-black hair, all one initially needed was the identity of the mother, the age of the infant and its sex. As Dieter/Dierdre had taught me, at that age sex was easy to assign: male baboons are born with penises about the length and prominence of a leg or arm.

It was harder to identify the large number of two- to four-year-old juveniles, let alone tell what sex they were. The giant penises of the baby males seemed to disappear entirely, leaving only the pattern of the hardened sitting pads (ischial callosities) as a way to distinguish between male and female. Males have an unbroken pad with a slightly raised center line below their tails, while females have a small space between the pads for the vaginal opening. I was to spend a lot of time during the following weeks trying to get a good clear look at bottoms. As with the adults, it was the tails of the juveniles that finally allowed me to begin to identify individuals. Crooked tails, chopped-off tails, tails in the shape of letters of the alphabet, bushy tails and skinny tails—the variety of subtle differences was astounding.

While I was learning to recognize individuals, I was also beginning to distinguish between what at first seemed like myriad behaviors. Fortunately, many of these were already well defined in the baboon literature. A female "presents" to a male when she approaches and turns | *greeting* her bottom toward his face; then the male will generally sniff her to see | if she is sexually receptive. Two males can also present to each other with no sexual intentions at all, and the same greeting is frequently exchanged by females, juveniles and even babies when they can manage it. Making a list of behaviors, called an ethogram, is always somewhat tedious, and in this case it was also comical, because I had to struggle to make "human" words describe baboon behavior as accurately as possible.

I encountered plenty of challenges. What did the look Ray was giving Sumner mean? When Sumner pushed Rad away from a Grewia tree laden with fruit, Rad's eyes seemed to shoot lightning; he'd raised his brows and opened his eyes very wide so that the sun caused his bright white eyelids to flash. It was a devastating look, utterly clear in its meaning. Was this what others called an "open-eye threat"? After much fumbling around, I labeled it an "eyelid flash"—neat, descriptive and with no anthropomorphic interpretation possible. I could decide later whether the two terms referred to the same thing.

*threat/ anger* |

Now that I could recognize some individuals and distinguish between some behaviors, it was easier to make sense of baboon interaction. Although I was beginning to figure out which babies belonged to which mothers, all the older infants and juveniles also obviously had families, as did the adolescent males, females and adult males and females. Aside from the adult and some adolescent males who came from other groups, everyone was related; I would need to piece together how many families there were, and who belonged to which.

In order to do this, I had to pay close attention to the preferences each baboon exhibited. Communication was critical. Every aspect of body language seemed fraught with meaning—baboon communication for the most part conveyed the *emotional* state of the actors: where they stood, how they held their bodies, what they did with eyes, hands and tails—everything was important. And, in turn, emotions told me about relationships. There were friendly greetings and aggressive threats. Gradually I began to perceive associations, but I knew that every hunch would have to be confirmed over and over again. Did juvenile male Sherlock groom the adult female Anne regularly? Was he protective toward Andy, her infant? Did he rest, walk, feed and sleep with her primarily? Exclusively? If so, there was a chance that they *were* related. I started to reconstruct hypothetical "families." It was like watching silent movies without the words being flashed on the screen to guide you, following each plot uncertainly at first, then with growing comprehension.

Today, though, after my adventure out of the van, I was still intent on the most basic of basics. The sun was directly overhead, and as I sweated, it gradually descended to the horizon. I had seldom concentrated for so long; my body and brain ached, but I had stayed alert, both to the dangers Matt had warned me about and in expectation of Ray's inevitable attack on the males of Pumphouse. When would he make his move?

The light started to fade; the troop moved in little clusters toward the sleeping cliffs. Ray plodded along far behind; we'd both spent the day in more or less the same fashion, each keeping a polite distance from the troop.

Ray was still with Pumphouse the next day. He again assumed a position that looked like that of an orchestra director, with his devoted musicians in a semicircle around him. Today, and for many days to come, he was the observed outcast, whose aloof presence was soon taken for granted.

---

As I relaxed about Ray, no longer anticipating his aggression, I began to feel easier about my own role. Nothing I did was done rapidly; the whole process took several months. But gradually I was moving closer, and finally, on foot, into the troop itself. I was fortunate that I had started my study in the season of short rains; life for the animals was lazy and luxurious, and the physical demands on me were minimal. The baboons traveled very little; they spent most of the early morning relaxing on the cliffs, grooming, playing, socializing. By midmorning one of them would initiate a descent to the grassland, and feeding would begin.

At midday, the troop would wander off to a watering site a mile or so away. I would follow at a safe distance, as would Ray. After an interval of drinking, resting, grooming and playing, we all returned to another good feeding spot, where Ray would resume his position in the semicircle.

The pace was easy. Weeks and then months passed; seasons changed; Ray still played the role of benevolent outcast. Tension did seem to be easing slightly, but any initiative taken was on the part of the troop, not him. Occasionally a young female had ventured near him, tail erect in a "present" posture. All these adolescent females were sexually receptive, but the adult males of Pumphouse didn't seem to find either them or their tight red genital swellings attractive. They preferred the softer, pale ones of the adult females. The young females sought partners anywhere they could, but all they found were juveniles and sometimes even infants. When one approached, Ray would shake his head from side to side, pull his chin back and narrow his eyes, smacking his lips. Tim Ransom had called this a "come-hither look," and I adopted the phrase for my ethogram. But once Ray had inspected the adolescent's swelling, he sat still, waiting for her to make the next move. One or another of them would occasionally work up the courage to groom him, which he seemed to enjoy immensely, relaxing in total abandon, every muscle limp except for that part of his body held out to be groomed. But all these episodes were brief. The females weren't relaxed and would return quickly to the rest of the troop when Ray showed no more sexual interest in them than had the other males.

*tension*  When Ray finally moved, it was sudden and in a direction I would never have predicted. He approached a female, Naomi, on the edge of the troop—the one I had spotted earlier from the van because she, too, seemed like an outcast. Ray walked toward her slowly and determinedly. Naomi seemed to take a few seconds to collect her wits, and I wondered what was going through her mind.

I watched, spellbound. Naomi tensed at Ray's approach, her tail up in the air, presenting to him. Her nervous glances both at and away from him quickened as he came nearer. At the last moment, it was all too much for her. She uttered a low scream of fear, crouched slightly and ran off, tail still erect, her mouth in a "fear" grimace.

Ray, the picture of benevolence, strode after her. Having watched him for months, I realized that his swagger was designed to convey as much goodwill as is possible for an adult male baboon to demonstrate. Again he gave his come-hither look, the one he'd been practicing on the little adolescents, and came nearer, still smacking his lips, pulling in his chin and narrowing his eyes. He had never had a chance to sniff Naomi's bottom, but he and I could both tell that she was not sexually receptive, just by looking at the flat skin where a swelling would have been. So what could he possibly want with her? During his next approach, he added a low grunt to his other gestures. Naomi dashed off. Again he tried, and again failed. Giving up, he resumed his Buddha-like posture, but this time much closer to Naomi than ever before. He was watching *her*, not the rest of the troop, and so was I.

Naomi was not the only one disturbed by the change in Ray's behavior. As far as I could tell, she had two children, a year-old daughter, Robin, and a juvenile son, Tim. Robin was already old enough to spend much of the day with her playmates and other friends, but she frequently returned to Naomi for reassurance when the going got rough, or to sit quietly enjoying the contact of her mother's body and hoping to be groomed.

Now, after Ray's campaign, Robin ran blithely over to Naomi, not expecting any problems and so not at first noticing the subtle shift in Naomi's posture as she sat feeding. Ordinarily Robin would have tried to crawl into her mother's lap and hope to suckle at her nipple, but suddenly she stopped short; something was wrong.

As she peeked around Naomi's body, Robin caught sight of Ray feeding quietly nearby. He was *close*! The yearling was startled and jumped back, frightening Naomi, who quickly stood up to locate the trouble. Following Robin's glance, she concluded that the infant had been upset by Ray; she wasn't keen on Ray's presence, either, but she'd begun to relax a little, even though he'd been following her for only about an hour. Relieved that this was all, Naomi settled down to feeding again, making certain to place her back to Ray so that she was effectively blocking out his presence in the only way she could. Robin, still confused by the situation, glanced quizzically at her mother, then at Ray. When she realized that her mother, though tense, was not frightened, she relaxed and a few minutes later was sleeping peacefully in her mother's lap, her small head lolling backward and Naomi's nipple held between her lips.

A similar scene occurred with Tim, who was a fully weaned juvenile, and considerably more independent than Robin; he would often come to sit with his mother and sister during the day. Tim noticed Ray much sooner than Robin had. He seemed caught between conflicting emotions, for he clearly wanted to be with his family, perhaps hoping that Naomi would groom him while Robin was asleep. He hadn't ventured this close to Ray since the strange adult male had first appeared. When this had happened, months earlier, Tim had run to greet him, panicked and rushed off to watch him from a safe distance. Now he approached his mother cautiously, trying to ignore Ray, and offered his side to be groomed. Naomi didn't respond and continued to tuck handfuls of grass shoots into her mouth. Tim edged in closer, blocking the route between the ground and her mouth. Determined not to have anything interfere with her feeding, Naomi slid to one side and Tim began grooming Naomi, hoping that she might reciprocate.

In the days that followed, Naomi had a constant shadow. At first this made her nervous, and she would frequently glance up from whatever she was doing to check on Ray, who patiently followed her week after week, keeping his distance and sitting most politely still. Soon, his deference began to be rewarded: more and more of the troop appeared relaxed around him. But they still did not interact. By now, Ray was more like a familiar piece of the landscape than a regular member of the Pumphouse Gang.

Then, in the space of two days, Ray transformed his world. It began without any fanfare—just another episode of Ray following Naomi. But this time he didn't stop at his usual acceptable distance. Walking in a friendly manner, grunting, lipsmacking and sending out as many appeasing signals as he could, he invaded Naomi's own personal space. She was startled, but seemed more cautious than scared. Why was Ray approaching her after all these weeks? What did he want? Would she return his friendly signals? How did she really feel about him? Her body language seemed to communicate considerable ambivalence.

By this time, Ray was within arm's reach; Naomi stood up slowly and presented her bottom to him. When Ray reached out to touch her, she pulled back, but sat down barely twenty feet away, waiting and watching. They exchanged glances. Ray showed some signs of nervousness too, as he checked the reactions of the nearby baboons. No one was watching; no one seemed in the least bit interested in the drama playing itself out nearby. He tried again, and then again. He received little encouragement, but Naomi neither screamed nor fled very far. At the end of their fifteen-minute duet, Ray was sitting much closer to Naomi than ever before. He seemed satisfied with his success, at least for the moment.

The next day, as if emboldened by the previous day's victory, Ray went even further. He had been trying to touch Naomi for months, but when his first efforts had so alarmed her he had contented himself with sitting as close to her as possible. This time, Naomi's reaction was simply wary, and Ray took this moment to press home. She pulled away from his approaches much more slowly, and Ray finally began a grooming bout. Both partners were nervous, and their tension created a strangely amusing scene. Grooming partners are usually the epitome of relaxed bliss, the groomed individual appearing transported to another world, giving in to the pleasure of contact and trusting that the groomer will be vigilant to any dangers. Although working hard at the task, the groomer is also relaxed. The perfect peace emblem should not be a dove, but two baboons grooming. Yet Ray and Naomi were responding to their mutual contact at two conflicting levels; the grooming itself had pushed all the right relaxation buttons, yet the unexpected contact between two individuals who were still wary of each other created tension. Naomi was having difficulty staying awake, unconsciously falling into the reverie of a grooming stupor and then starting at some imagined danger. Ray was an emotional yo-yo, sensitive to Naomi's every shift in muscle tension, riding the roller coaster of relaxation and vigilance along with her. Despite Naomi's initial reluctance, it seemed that she had come to understand Ray during the past weeks, even though there had been no direct contact between them.

I was discovering that for baboons what doesn't happen is almost as important as what does. The fact that Ray hadn't bullied Naomi, hadn't acted aggressively to the animals around her, was persistent yet patient and—most important of all—had produced the correct friendly gestures at the right times had allayed some of her fears.

Eager as Ray had been for Naomi's attentions, after that first day of acceptance it was she who became Ray's shadow. His friendship with Naomi pulled him into her family circle. He could sit close to Robin when she was near her mother, and even Tim adjusted to having him around. Ray's fortunes had taken a radical turn for the better; the friendship had changed his status with the rest of the troop, too. With Naomi's stamp of approval on him, other baboons were less apprehensive.

*newcomer → violent*

I was pleased for Ray, but also confused. I had expected him, as a new male within the troop, to start out by being aggressive, to fight his way into the middle of the group the first day, to grab what he wanted and take over by force; if he couldn't do this, I thought he would leave and try to infiltrate another troop. I had anticipated his aggression from the beginning, and was surprised when he simply sat and watched. I was surprised again when his first move was to follow Naomi, and even more startled when he finessed his way into her affections rather than forcing her to submit, which he could easily have done by virtue of his larger size and strength. After months of waiting for the explosion, I finally breathed normally again.

Now I had a lot of questions. Ray was as big as any of the Pumphouse males; why didn't he fight them? Why should he care so much about being friends with Naomi? Why should Naomi be so slow to accept his advances yet, once committed, be as interested in Ray as he was in her? When would the other males notice Ray? If males were built as fighting machines—and they certainly seemed to be—when did they actually fight? Was this really as peaceful a society as it seemed? If so, how was the peace created?

Ray was not content with his newly won place in the group's social sphere; he wanted more. He approached a number of other females to test their reactions, but most of them were wary, even though they showed nothing like the exaggerated response he'd elicited during his first days with the troop.

I have no idea how Ray came to his final decision, but the next baboon he decided to pursue was Peggy. She was the highest-ranking female in the troop, and the one with the largest family. She played a special role in the group's daily life; she also came to play a special role in my understanding of baboons, because she was one of the first animals who permitted me to get close to her. Until her death in 1982, aged about thirty-two, she was my best guide to the intricacies of baboon life. Today, Peggy's skull rests on my windowsill—not a gruesome reminder of her death, but a meaningful symbol of her importance to my work and life.

Peggy was in her prime at the time Ray showed an interest in her. She was among the easiest of all the females for me to identify because there was something unique about every aspect of her. Her tail was smooth and straight, forming a ninety-degree angle at the point where it bent away from her body. None of the others had quite that angle or quite that shape. Her long face, too, was very distinctive: she had a black V-shaped patch of hair on the top of her head, a very pointed nose and a cataract that almost completely covered her left pupil. I wondered how this might affect her vision, but I discovered only two clues indicating that her sight was impaired. Whatever she found when she groomed another baboon she would examine carefully by holding it close to her good eye. No other baboon did this. And when she examined an infant, she would lift it high off the ground in one hand and then lower it to her upturned face, examining it with her right eye. This always amused me, but I was never sure whether she was actually investigating the infant or whether she just liked to play with it this way.

Ray couldn't have picked a better female to befriend—or a worse one. Peggy also had close friendships with two other males, both of whom had been residents of Pumphouse, at least from the time when Bob Harding had begun to watch them. Naomi, too, had had another male friend, Rad, while Ray was courting her. But although Rad watched with interest to see how the relationship would develop, he simply shifted his attentions from Naomi to another of his female friends and offered no objections when Ray finally made his move. Rad and Ray seemed to be avoiding each other. The rest of the males also ignored and generally refused to acknowledge Ray's presence in the group once the first days of tense greetings were over.

Ray's interest in Peggy did rouse her friend Sumner. He had a very distinctive face, and a tail that had been broken in one place; part of it had been lost somewhere, so that what was left looked like a square upside-down J. He had a mole on his chin, and his muzzle was also very square, appearing somewhat soft because his canines were either worn down or broken. He was a chunky male with a deep chest, but I was never certain how much of his bulk was muscle and how much was hair.

Sumner and Peggy were nearly inseparable. If I couldn't find Peggy, I'd look for Sumner, because, more often than not, she would be close by. I was still inexperienced as a baboon watcher, but even so I could notice a difference in behavior between Ray and Sumner. Sumner seemed better at all the "social" things; his actions were more adept than Ray's. They might both use the same signals, but Sumner's were invariably more effective than Ray's.

It soon became obvious to Ray—and to me—that Sumner was not at all pleased about Ray's intentions. He was not going to stand around and watch while Ray made advances to Peggy, and from the moment he became aware of what Ray wanted, he guarded Peggy jealously. If he spotted Ray heading in her direction, he would rush across whatever distance separated them, hurrying to her side, placing himself between the two and sometimes even chasing Peggy away from Ray if he got too close. *jealousy*

Ray tried again and again. I was fascinated, not just because of the drama, but because I simply couldn't understand what the males were doing. I could describe what *was* happening, and after a while I could predict what was *about* to happen. For example, I might be watching Ray, and at the point when he decided to approach Peggy again I would glance around to find Sumner, expecting him to be tracking Ray closely; he was. When Ray got within a certain distance of Peggy, I would predict that Sumner would act, and he always did. But I didn't understand why. Ray was certainly as big as Sumner, younger and in better health; why didn't he either take Peggy by force or stand up to Sumner? I was particularly confused because Sumner actually appeared *?*

to be frightened of Ray, as did all the other Pumphouse males. What was Peggy's role in all this? She seemed willing to comply with Sumner's wishes, yet she didn't avoid Ray's advances on her own, and occasionally she seemed to approach Ray deliberately herself.

Tension was mounting. Sumner became more nervous as Ray persisted, and the two males began to be more openly aggressive. It was a low-keyed aggression at first, with Ray opening his mouth in a wide, canine-displaying yawn—a typical threatening gesture—at Sumner when he interrupted Ray's move toward Peggy. At this point Sumner would rush at Peggy and begin nudging her away from Ray just by staying very close behind her and keeping moving himself. Ray would follow. When they stopped to rest, he would stop too, and repeat his threats. Staring, yawning, then lifting his head and exposing his canines again, Ray would add an eyelid flash to the yawn. Sumner turned and stood his ground. Ray was an impressive sight, hair fully erect, ferocious-looking mouth wide open, but he, too, was getting nervous, scratching his chest and glancing around even while sending out his threatening signals.

All this dragged on for minutes that seemed like an eternity to me—and probably to Sumner as well. There was a brief chase by Ray, and an equally brief counterchase by Sumner. Then things happened fast. The two males grappled, circling each other, pant-grunting. Sumner screamed—an adult male scream is a serious expression of fear—and it was all over as suddenly as it had begun. Neither male was hurt, and I couldn't tell who'd won. Perhaps Sumner and Ray were equally confused, since both walked off in opposite directions, leaving Peggy alone.

I found myself stunned and shaking. The aggressive signals Sumner and Ray had exchanged had a powerful effect on me, my humanness and scientific objectivity notwithstanding. Though the fight had posed no real danger to me, the adrenaline surged through my blood as if I had been physically involved. I calmed down after a few moments and reflected on the events. I could understand why Ray and Sumner had fought, but why now? Why not earlier or later? Why hadn't the winner claimed Peggy, and why wasn't anyone hurt?

---

By this time I had slowly, almost imperceptibly—rather like Ray himself—moved into the midst of the troop. I was always careful about how I moved and where I stepped, making sure it wasn't on an infant, who could be sitting half hidden in the grass. I was also always careful at whom I looked; when I came face to face with one of the animals, I'd lower my eyes or turn away my head. It was for this reason that I was unable to wear sunglasses, despite the wind and the blinding light; the first animals to catch a glimpse of me wearing them ran off in obvious terror. Small wonder: to them, the glasses not only covered my own eyes, obliterating important visual communication, but presented them with the biggest, wide-eyed threat they'd ever seen. Above all, I had to be especially careful about exactly where I interposed my body— never between two animals in any kind of interaction. To some degree it had been Naomi who'd taught me many of these things, for I had used her to gain entry into the troop. Poor Naomi: first Ray, then me. But she took it in her stride as long as I stayed well behind her.

After Naomi, I followed other individuals, always being careful not to press anyone too closely. I had been warned by other field workers to dress drably and always to wear the same clothes. This was more of a problem than I had anticipated. We were two degrees south of the equator, at an altitude of nearly 7,000 feet. The wind brought some relief, but added its own burden to my already abused skin. Since the temperature ranged from frigid in the shadow of the sleeping cliffs in the early morning to blistering in the midday sun, I had to dress in layers and peel off. Tank top covered by turtleneck covered by sweat-shirt made a bulky but practical outfit. There was no problem with the sweatshirt, but my underlayers were by chance of different colors; I couldn't wear the same dirty clothes day after day.

The baboons seemed not to care what I wore, what color it was or how different I looked from one day to the next, as long as I peeled the layers off slowly, not scaring them by any abrupt movements. They knew me by now, and I shouldn't have been surprised that they did, considering that they could recognize different makes of vehicles from as far as a mile away. They always ran from the Coles' white Peugot but not from that of the neighbors, which, except for being a year newer, was identical. I found out later that the Coles sometimes trans-ported ridgebacks—big Rhodesian hunting dogs—in their pickup, and the dogs occasionally jumped out and attacked the baboons. If the animals could make this kind of distinction, they could certainly tell who I was when I was standing in their midst, even if they had never seen a particular T-shirt before.

So, pink-nosed and squinting, I became the intrepid baboon watcher, going wherever the troop went, juggling tape recorder, clipboard and binoculars—observing, taking notes, thinking. I finally became comfortable enough to attempt a feat I'd been contemplating for days. It is not only baboons that have to answer the call of nature. At first I'd retreat behind the VW to relieve myself, but now the van was often miles away, and I hated to leave and miss something.

I decided I would pee on the spot. Trying not to move too quickly, I lowered my shorts. So far, so good. Suddenly every baboon around stopped dead in its tracks. They stared at me in wonder as the sound reached them. I thought I understood what was going on; up until then, I'd arrived, watched and left. They hadn't seen me eat, rest, drink or sleep. They hadn't been fooled into thinking I was a baboon. They knew I was human, but they had never been so close to one before and maybe they thought humans didn't have to pee. They stared, but not one ran away, and when I pulled up my shorts, they lost interest. The next time, they didn't react at all. As the baboons relaxed around me, I gradually relaxed around them. But even so, I was determined to preserve my role as a nonentity within the troop, never to interact with the animals, to be simultaneously tolerated but unobtrusive. It was natural that I found myself constantly wanting to reach out to them, to touch an infant, return the play invitation of a juvenile, groom a new baby, but to interact would change the nature of the study. Although I could learn a lot about the animals from their behavior toward humans, the approach that I championed insisted that the most important insights would come if their human observer always let the baboons be baboons, allowing them to remain as unaffected in their life with one another as possible.

I must admit it was a little different with Ray. I suppose that in his eyes, as in mine, we had a special relationship. We had both entered the troop at the same time, with the same patience and caution. Certainly our acceptance within it showed a number of parallels. One day, Ray surprised me. Socially speaking, he was at the end of a rather long losing streak. I was dutifully recording his bluffing interaction with Big Sam and Sumner when, without warning, he turned and rushed at me. I was more baffled than alarmed; none of the baboons had ever been aggressive toward me, even when they were being aggressive with one another. This, I had reasoned, was another benefit of not interacting with the troop: I was neither friend nor foe.

Yet here was Ray, making straight for me. There was no mistaking what he meant, and it took me only a few long seconds to figure out what was happening. Ray wasn't threatening me; he was soliciting my help. He wanted me to support him against the other males!

I checked his gestures quickly to make sure. There was no doubt about it: he was slapping the ground with his hand, looking first at me, then staring at the two males, then back at me to see what I was going to do.

This handsome, powerful male struggling to become part of the troop spoke strongly to something in me. I badly wanted to help him, but I couldn't. I signaled this by turning away completely, leaving Ray to handle the situation on his own. Suddenly I realized how far I had come from those early days at the Red House. I was tougher, leaner, braver, but the biggest change had been in my attitude toward the animals. Ray won his struggle alone, but I shall never forget how honored I felt by the compliment he paid me.

The day ended as so many had ended before, with the fading light signaling that it was time for the animals to return to their sleeping cliffs. There was nothing different about the day; *I* was different. My mind was full of new questions and my heart of new emotions.

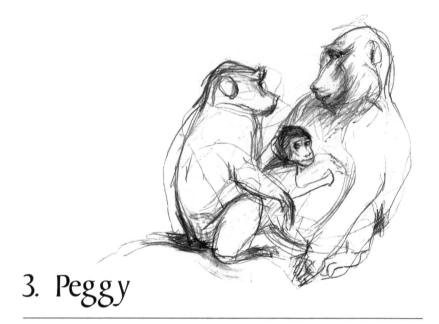

# 3. Peggy

It had taken months, but it had finally happened: I could wander through the troop at will, seemingly invisible. I could identify each baboon at a glance. Now I was ready for the real heart of my study: obtaining detailed information on every individual, and determining the relationships between and among them. Like Ray, I needed a special guide to help me understand troop dynamics; like Ray, I chose Peggy.

Peggy was important. In disputes, she could get whatever she wanted—a tidbit, a comfortable spot in the shade—simply by moving in and waiting. She was the highest-ranking female in the troop, and her presence often turned the tide in favor of the animal she sponsored. While every adult male outranked her by sheer size and physical strength, she exerted considerable social pressure on each member of the troop. Her family also outranked all the others, a fact that took me months to clarify. I was fortunate in that Peggy's family was easy to identify because they all resembled her. All were deep brunette in color, and shared the same shape of face, angle of eyebrows and eyes, despite the fact that probably each had a different father. And the tendency of Peggy's family to spend time together was as striking as their physical resemblance. Even before Pebbles, her youngest, was born, Peggy was usually accompanied by one or more of her children or grandchildren, grooming and being groomed, while the others played nearby. If an outsider threatened or frightened one of them, they all came alert in a quick show of solidarity.

As I saw here and confirmed frequently, the mother or matriarch gives most of her time, attention and protection to the youngest member of the family. Older children get shortchanged unless they consider their requests carefully. Consequently, a rank order develops within each family: the mother is dominant, the youngest child next, then the next youngest, and so on, in reverse birth order. What begins in the bosom of the family and depends initially on the mother's intervention soon develops into a system that operates without her support and persists into adulthood.

In Peggy's family, sibling rivalry did not seem as intense as in other families. This was partly because of age and spacing; Peggy was an older female, and her reproductive rate had slowed down. The age gaps between her children—Thea, the adult; Paul, the adolescent; Patrick, the juvenile—were much greater than average. The gaps could have represented children who didn't survive, or who had migrated out of the troop or simply Peggy's lowered fertility rate.

I also felt that another reason for the contentment in this particular family was Peggy's personality. She was a strong, calm, social animal, self-assured yet not pushy, forceful yet not tyrannical. Compared with some of the other females—notably her own daughter, Thea—Peggy was socially brilliant.

When I first began observing the troop, Patrick was a yearling and *mother-son* Pebbles had not yet been born. Peggy gave Patrick priority over Paul and Thea, allowing him to nurse, ride on her back and sleep in her lap. This didn't seem to bother Paul in the least. He was now larger than his mother, and though dominant to every other female in the troop, maintained his subordinate rank to Peggy. It was a tranquil mother-and-son relationship, marked by mutuality. Peggy rarely intervened when Paul was working out his relations with other males, but if he was really in trouble she would come to his support, often placing herself in real danger. When this happened, it actually seemed as if she turned the tide for Paul, not making him a winner but preventing him from losing badly.

Paul returned the favor—not frequently, but at critical times. Peggy was already dominant to all the females in the troop, so her interactions with them were more or less determined in advance. Although it was not in her character to claim the prerogatives of rank as often as she might, when she really wanted something she got her way sooner or later. This was not the case with the large adult males in the troop, who would sometimes confront her. The support of her adolescent son could not turn defeat into victory on these occasions, but at least it tempered the nature of the defeat. Paul was risking a great deal in supporting his mother; he could not singlehandedly take on and defeat any of the troop's adult males, but he certainly tried.

*mother-daughter* Thea's relationship to Peggy was another story. Sometimes she sought out her mother's company, making Peggy her preferred grooming partner, while at others she seemed to avoid or so ignore her mother that it was difficult to tell they were a mother-daughter pair. But no matter whether Thea was ignoring her mother or fawning upon her, she definitely relied on Peggy when she got into difficult situations.

Thea was, in fact, a bitch. Her status in the troop was second only to her mother's, and she used it tyrannically; she was unprovokedly aggressive, intimidating other females in situations where Peggy would have calmly quelled the whole matter with a rebuking glance or approached and waited for what she wanted. Moreover, Thea was always poking her nose into other people's business. Whenever females or juveniles were involved in a tiff, Thea would be there in seconds, adding her weight sometimes to one side, sometimes to the other, frequently almost schizophrenically switching sides unpredictably. She often managed to *prevent* the quarreling individuals from settling the argument.

When the situation got this far, Peggy would occasionally come to Thea's support, her added weight resolving the issue on the spot and cutting short Thea's muddling influence. Most of the troop gave Thea a wide berth, so these situations occurred less frequently than they might have, but at times she could stir up such a to-do that some of the adult males would intervene and she would find herself the new underdog. Here, as with Paul, Peggy would lend her support to her daughter, usually tipping the balance in the direction of a face-saving retreat on Thea's part.

I badly wanted to understand Thea's ambivalence toward her

mother. What was she thinking and feeling? What was driving her? What made her so different from Peggy in her interaction with the other females? Observation could take me only so far: what I really needed was an interview!

*grandmother – children*

Thea's children, Tessa, Theodora and Thelma, were as positive about their grandmother as Thea was ambivalent. In part, Thea's character traits may have encouraged this. Her lack of calm and her generally high level of aggression sometimes extended to her own children; certainly she seemed to have less time for her daughters than did other mothers.

All three granddaughters sought out Peggy, who was so generous with her grooming time that it sometimes appeared as if she had an incredible number of children and that Thea had not even begun her own reproductive life. The dominance rank within Peggy's family was stable and predictable: Peggy was at the top, with Paul as her equal; Patrick next, then Thea. Of Thea's children, Thelma was on top—as the youngest, she was most often on the receiving end of Thea's erratic aid—and then Theodora; Tessa, the oldest, was at the bottom.

*family hierarchy (Peggy)*

As far as the rest of the troop was concerned, it didn't matter with whom one grappled. Each member of Peggy's family was accorded the same high status as Peggy herself; this made sense because when things got rough for *anyone* in her family, Peggy would eventually pitch in herself.

––––––––

Birth brought changes into this stable family structure. Years later, when I had my own baby, I saw new similarities between baboons and humans. For both, an enchanting, demanding, helpless newcomer can make all the difference to an established relationship.

I, as well as Peggy, had been expecting Pebbles's arrival for six months following the cessation of Peggy's sexual cycle. When she was about six weeks pregnant, the bare, pinkish skin on her bottom began to deepen in color (because of her age, her bottom, unlike the gray ones of the young females, was always slightly rosy), first becoming petunia-like, then a deep scarlet. It was from observing these color changes that I could establish the time of the baby's arrival: approximately 160 days from the last appearance of Peggy's full-blown sexual swelling. During these months I could also decide on a name for the baby. I gave each new baby a name beginning with the first letter of its mother's name. I would always select two names, of course—one for a male and one for a female, leaving the final choice until I had determined the infant's sex.

Sure enough, a mere two days after the due date I found Peggy off by herself; she was a little more wary and secretive than usual. And there on her belly was Pebbles. She was covered with jet-black hair, from which a tiny bright pink face and ears poked out. Her wet, limp umbilical cord still hung from her, and she looked crinkled and damp all over.

Calm and confident as usual, Peggy trusted me to get as close as she used to before the birth, which was unlike many of the baboon mothers, who were so possessive that it was sometimes weeks before I could even tell the sex of an infant.

Like all newborn primates, Pebbles looked out of proportion. Her head was much too large for her body, and her limbs were spidery. But she was far more developed than a human baby; she could hold her head erect, could cling to her mother with only a little additional support, and within a week could toddle around unsteadily on her own. Most of the time she was attached to Peggy, which meant that at first Peggy had to hobble along on three limbs, supporting the baby with one arm. This necessitated slower travel and frequent rests. Pebbles clung on not only with her own grasping fists and with the help of Peggy's arm; she was stuck firmly to Peggy's nipple all the time. If a sudden movement dislodged it, she would rub her face from side to side and up and down, rooting around until she found it again. Peggy's nipples were long and tough from years of nursing; they never seemed to have the sore appearance the nipples of the younger mothers had. A female nursing her first baby usually had pink, swollen, budlike nipples, and would often massage them as if they hurt.

For the first day of Pebbles's life, her eyes remained tightly closed and she made only a few sucking noises, mixed with coos, quiet grunts and small groans. Her loudest sounds came when she was in distress: Peggy sometimes loosened her supporting arm, and the infant lurched back dangerously. Her pitiful screams and moans alerted her mother, who scooped up the infant and nestled it safely in her lap. Everyone nearby participated in the crisis, turning to watch until the baby was righted again and then giving reassuring grunts and lipsmacks.

Indeed, the entire troop was fascinated by Pebbles. Several females approached, attracted to the infant but a little wary of Peggy's high rank. It was further proof of Peggy's benevolence that they did try to make contact. When Thea gave birth to a new infant, few outside her

T. W. Ransom

1

1. Kekopey Ranch, in the middle of Kenya's Great Rift Valley

2. A male shows his impressive canines

T. W. Ransom

2

3. Ray was a key figure in understanding the males

4. Big Sam copulates with Peggy

5. Infants start out riding on their mothers' bellies: here, Benjy takes a peek from underneath Beth

T. W. Ransom

3

T. W. Ransom

4

T. W. Ransom

5

S.C.S

6

T. W. Ransom/National Geographic

T. W. Ransom

8

6. Male and female feeding

7. Baboon on sleeping cliff

8. Males and females differ
impressively

T. W. Ransom

T. W. Ransom

9. Mother and baby drinking from a rain pool

10. Juvenile feeding on acacia

11. Naomi with dead Hal

12. Pumphouse resident Rad with infant friends

S.C.S.

T. W. Ransom

10

11

12

T. W. Ransom

13

T. W. Ransom/National Geographic

14

13. Two black infants playing

14. Harriet tries to take Naomi's infant, Nanci

15. Male aggression—spectacular but rare

16. Infant riding on mother's back

T. W. Ransom

15

T.W. Ransom

S.C.S.

17

17. Black infant

18. Peggy shares Sumner's
gazelle

T.W. Ransom

18

family dared to come close. But Peggy, genial as always, accepted these advances while continuing to devote her attention to Pebbles. After presenting to Peggy, lifting their rear ends in the new mother's face, the females would run around and crouch down at Pebbles's level, eyeing the newborn, grunting softly, smacking lips and reaching out hesitantly, hoping for a chance to touch the baby. This tentatively accomplished, they would retire to a nearby spot, awaiting another opportunity to approach. Only Constance, Peggy's close friend, ventured to groom Peggy on the first day of Pebbles's life.

I was eager to see how Peggy's family would react to the new arrival, Thea especially. Patrick was the first to appear. As the displaced youngest, he had the most to lose. Wary and curious, he circled Peggy and the baby. His mother didn't even glance up; he circled again and again, looking first at Peggy, then at the peculiar new object in her lap. Finally he settled down nearby, watching and waiting. A little later Paul showed up, a self-sufficient equal. He took in the situation at once, gave Pebbles an avuncular grunt and sat down next to Peggy, grooming her.

One by one, Peggy's grandchildren appeared. They seemed to be shocked to find their grandmother, who in some ways was their surrogate mother, totally involved with a new rival. Tessa and Theadora, the two eldest, showed the same interest the other females had, trying to get close enough to touch Pebbles. Again Peggy was permissive, allowing repeated contact. Thelma, the youngest grandchild, seemed the most disturbed by the new arrival, approaching and withdrawing repeatedly, as if caught between attraction and jealousy.

Last of all came Thea. She marched right up to Peggy, showing none of the deference exhibited by the other adult females. Peggy in turn showed none of her usual affability and ignored her daughter. Thea sat very close, staring hard at Pebbles.

By the end of the morning, Peggy's entire family had paid their respects, each in his or her own way, characterized by age, rank and personal traits. The youngest seemed the most upset, the oldest the least troubled.

After the first few days, as Pebbles became more active, the troop began to focus intently on the new baby. Many young females diverted their routes so that they could look at, grunt to or perhaps touch the infant, and several of them frequently groomed Peggy, hoping in this way to have a better chance of being able to touch Pebbles. I was reminded of the visiting that follows the birth of a human, of how strong the impulse is to interact with the newborn. Studies have shown that the rounded faces and features of human babies elicit a "cute" response in human adults; does the tiny pink face and dark hair of a newborn baboon do the same?

I watched Peggy's family adjust to the newcomer. Patrick, who was nearly two, was to all intents and purposes self-sufficient, supplying his own food and getting around on his own. Before the baby arrived, Peggy had allowed him to suckle now and then, and would even let him ride on her back when the troop sensed danger. He still sometimes slept in her lap or nestled close against her on the sleeping cliffs. Now here was this little hairy object stuck to his mother, and everything had changed. Whenever Patrick came to be groomed, his mother was grooming Pebbles. He would put himself in a grooming posture, inviting Peggy to attend to him, which had always been enough before, but now she stayed intent on the infant and Patrick was ignored. Eventually he settled down very close and began grooming himself. There were no longer any chances of a suck at the nipple or a free ride, no longer a lap to nestle in. For several days Patrick tried hard to get Peggy's attention and reestablish the old relationship; he did everything he knew how to do, even interposing himself between mother and baby whenever he could, sometimes successfully but generally not. He adjusted gradually, eventually seeming comfortable in his role as older sibling and showing some interest in Pebbles. Finally, he sought out his older brother, Paul, as a substitute for Peggy, often sitting near him for comfort and to be groomed.

Peggy's youngest granddaughter, Thelma, had a more difficult time. She circled mother and baby warily for several days until she apparently decided that enough was enough and presented herself for grooming, just as Patrick had done. As with Patrick, Peggy was oblivious, but Thelma persisted and went on persisting. Every time Peggy looked up, there was Thelma, thrusting one part or another of her body into her grandmother's face.

Finally Peggy lost her temper. She grabbed Thelma firmly around the middle and gave her a quick nip with her front teeth. Thelma screamed. When Peggy released her she darted off, stopping a safe distance away to regain her composure, looking steadily at Peggy and Pebbles. After a while, although she continued to moan, first loudly and then quietly to herself, she sat gingerly grooming and apparently trying to understand this strange turn of events. Her grandmother had never handled her roughly before. She watched Patrick present to Peggy and be groomed. The moral seemed to be that Peggy had less time available, and her priorities had changed. Her first responsibility was Pebbles, then Patrick—and only after that, if she still had time, Thelma herself.

———————

I gained a better understanding of both family structure and troop organization from watching Peggy's family. Females and their kin were definitely the stable core of the troop. Males came, like Ray, and others went. In this way, baboons resembled rhesus macaques after all. Rhesus monkeys on Cayo Santiago Island, Puerto Rico, had been studied for several decades. Although they now lived in a strange environment, far from their native Asia, they were still free to roam their island home as naturally as possible. Both the Cayo macaques and the Pumphouse baboon kin groups, or matrilines, are headed by an older female, and include all her offspring and their offspring. Matrilines in Pumphouse spanned three generations, as far as I could determine—perhaps more—and extended sideways to include aunts and cousins.

These female-based units were the core of the troop, and it was the females and their families, not the adult males, who could be ranked linearly into a stable dominance hierarchy. Every member of Peggy's family outranked every other member of every other matriline. Pebbles herself, a mere infant, would come to dominate a fully adult female such as Harriet. Peggy might have to intervene on Pebbles's behalf at first, but Harriet would soon catch on and after that Pebbles would stand up for herself all alone. Every family could be placed within the hierarchy in orderly fashion.

———————

In addition to teaching me about family and troop structure, watching Peggy's family taught me exactly how a baboon grew up. Once Pebbles became a bit steadier on her feet, she tottered away from Peggy to explore her environment. Everything was worth close inspection: rocks, twigs, a small hole in a bare piece of earth. The world was large for Pebbles, and full of extraordinary surprises, many of her own making. She'd set out for one destination only to arrive suddenly at another. Falling down, which she did constantly, brought her eye to eye with new curiosities, and her short attention span would soon cause her to forget her original goal. She couldn't wander far at first, and was more intent on watching what her mother was up to. Peggy never taught

Pebbles anything directly. For the moment she was an indulgent mother, content to let her daughter use her as a jungle gym, not objecting to Pebbles's fingers poking into her nose and eyes.

*learning infant + mother*

Pebbles's concentration on Peggy helped to teach the infant many things—some of the essential baboon skills, and where to find all the necessities of baboon life. It could be amusing. At one point, Peggy prepared a site where she could feed on corms, the small, bulblike storage parts of a sedge called oniongrass. Pebbles watched as her mother grabbed a tuft of sedge and pulled against it with her whole body, yanking up the grass to reveal the succulent corms. It looked easy. Pebbles reached for a tuft, pulled with all her might, landed on her bottom and stared in shocked surprise at the few broken blades of grass clutched in her small hands. Oh, well, she seemed to say, I wasn't really interested; she shook herself and toddled off.

Pebbles learned from some of the other baboons as well. Theadora and Thelma would play catch-as-catch-can around their grandmother, with Pebbles watching wide-eyed, not quite knowing what to make of it all. The first time Thelma loped over, mouth wide open, her lips hiding her teeth (the classic primate "play-face"), Pebbles was frightened. But Thelma gently rolled the baby to the ground and after a moment, as the chase resumed, Pebbles got up and toddled a short way, following the playmates.

After this had happened many times, with the players always approaching with the same "face," Pebbles came to associate this expression with certain behavior, and more important, with a certain attitude: "It's all in fun; nothing is serious. I may look aggressive, but I don't intend any harm." Pebbles tried hard to make the same face herself, even though she didn't have any teeth to cover and couldn't have hurt anyone if she'd tried.

Once in a while, play would get out of hand, and what had started as fun could change imperceptibly to a serious dispute. The older might have been too rough, the younger might not have cooperated in just the right way. It was at times like these that the soft chortles heard only in play would turn to real screams. Teeth would be bared and expressions would undergo a complete transformation.

Watching from the sidelines, Pebbles learned the difference between playful skirmishes and serious battles. She was too young to join in yet, and perhaps her black coloration indicated her infant status to the others and protected her from bullying, but the minute she *could* actively move around, she played. Play is an important activity for all

*play*

primates, and as I watched I learned as much as Pebbles. Her first play was solitary. Baboons don't have toys as such, but when they find something they can play with, it turns into one: trees become acrobatic bars, rocks become slides or hurdles, bushes are hiding places and holes are even better ones. Whether slowly exploring the world around her or racing at breakneck speed up and down a tree, Pebbles was learning. But for a bright and social creature such as a baboon, solitary play becomes boringly static. An unpredictable and ever-changing partner gives play new meaning and excitement.

Of course there were certain guidelines Pebbles had to follow: a tree cannot object to anything, but another monkey can. Play cannot be too rough or too noisy; play and fighting must not get confused. The games of the older youngsters—hide-and-seek, roll-and-tumble, catch-as-catch-can, skydive—seemed to have simple but specific instructions. Skydive was one of the most exciting: one player forces another so far out on a thin branch that he has either to jump or fall to the ground, often as far as twenty feet below, followed by his partner, who seems to want to land on top of him. I remembered a film I'd seen of the Cayo macaques. *Their* skydive was from fifty-foot trees into a shallow pond below; at just about the same moment the first player surfaced, the second landed on top of him, forcing them both underneath again.

Like true play in humans, as opposed to games, the object is not to win but to keep the play going, even if it means purposely missing your partner or being forced to climb back up into the same tree and edged into jumping from yet another branch.

---

Tessa, Peggy's oldest granddaughter, was also learning. She was about four and a half, and just approaching puberty. At first, her sexual swelling resembled a tight little pink bud instead of a large pale cushion. She retained the swelling much longer than an adult female would. This pattern was repeated over and over again, but each time the bud bloomed more and deflated sooner, until the size and timing, though not the color, were like that of an adult female; the color was still very bright, almost red.

Right from the first, Tessa acted as if she had a very appealing swelling. She would strut from one adult male to another, and brazenly or shyly, depending on the male, present her bottom. The males' lack of interest disappointed but did not deter her. After she had exhausted the adult males, she tried the adolescents and finally the juveniles. A few did approach, sniff and mount her, but these young males had not yet reached their own sexual maturity.

Were these practice copulations? Lab studies have shown that monkeys reared in isolation from their mothers, playmates or both are sexually incompetent as adults. Doing what comes naturally is apparently not natural after all. Even infants showed interest, and Tessa was patient with their often comical attempts. They did seem to improve with practice.

By now, Tessa was quite a sight. Her swelling was beginning to reach mammoth proportions, much larger than that of Pumphouse adults; in fact, it resembled those I'd seen in zoo animals, who often go through sexual cycle after sexual cycle and never become pregnant. One day Tessa presented to Strider, a male in his prime, just as she had a million times before. By now she was quite perfunctory about it, as if she knew what the response would be; she would sometimes even leave before the male had finished his inspection. I was as surprised as Tessa when Strider grabbed her; he *was* interested. Shocked, perhaps even frightened, Tessa wriggled free and put as much distance between them as she could manage. But Strider followed, eventually coming close enough to touch her. She stopped and looked at him while he took the opportunity to groom her: grooming helps to establish "ownership" in everyone else's view. Only after Tessa had calmed down a bit did he rise to mount her. It was an incomplete copulation, because Tessa was still nervous and pulled away too soon, but it was her first real "consort," as the sexual pairing of baboons is called. It looked as if Strider would have his work cut out.

Tessa would have to learn, too. Learning during infancy and adolescence was impressive: Pebbles was born knowing next to nothing. She had no idea of what to feed on, except her mother's milk; no idea of where to go, except her mother's belly; not even, it appears, how to reproduce. By the time Pebbles was as old as Tessa, she would know the ropes. But learning didn't stop at that point. Even for an adult, each day brought new challenges, both physical and social, new lessons that had to be learned if one was to become a successful baboon.

---

I reviewed what I had learned from watching Peggy. Life begins within the small, intimate world of the family: close, caring, protective. Nothing seems to match these family relationships, yet an individual's social horizons do extend beyond the family. There are consorts with the opposite sex for adults, but there is also something else—attachments between two unrelated animals, attachments that closely resemble family relationships. Unlike relationships within the family, friendship is based on nearly equal give-and-take. While children *expect* to be cared for and protected regardless of whether they return the favor, friends expect favors in return for favors. As I was to learn later, friendships were often long-lasting. Friends appeared simply to like being together, sitting, resting, sleeping or grooming. Such mutuality could become a complicated give-and-take ritual that extended over days or even weeks.

Peggy herself had such a friend: Constance. The two were almost inseparable. If I wanted to follow Peggy, I often looked for Constance because this doubled my chances of finding her friend. When I was first putting the Pumphouse families together, I wondered if Peggy and Constance were related, but I soon realized that their closeness in age ruled out the possibility of a mother-daughter pair, and that their relative dominance ranks—with Peggy older *and* dominant—ruled out the possibility that they were sisters. In any case, the physical resemblance was not strong enough to justify the idea. Peggy was dark, Constance medium-colored; although they both had long tails, Constance's fell in a gentle curve and Peggy's jutted out at an angle. The shape of their faces and the slant of their eyes differed, as did their general body gestalt. Over the years, as the relationship remained the same, I realized that Peggy and Constance followed the pattern of females I knew who were "friends," not relatives.

Friends of the same sex seemed to have a very stable relationship. They were grooming partners and companions. They supported each other within the troop—but only up to a point. Although Constance was high-ranking herself, Peggy, who outranked all the other females, was useful to her—except in a situation that involved a member of Peggy's family. Then Peggy would side with her own relative. In her turn, Constance aided Peggy, but her help seemed more a token repayment, since, as in the case of young Patrick, her support could never be the basis for a win. Still, she would sometimes tip the balance just enough so that Peggy didn't lose to an adult male.

Despite the stability of their friendship, there were several ups and downs over the years in Peggy and Constance's relationship. They were so close that their families grew up together, but because the two females frequently had children at approximately the same time, family demands did impinge on friendship. When Peggy's family happened to be especially large, she had less time for Constance; only so much socializing could be done in a day, and this had to be divided between family and friends. When the family structure changed, Peggy's friendship with Constance took up the slack in her social time.

Constance did not bear grudges. It was not as if she had been rejected. The two friends had a gentle and finely tuned relationship that permitted the correct adjustments.

Females were not the only candidates for friendship. Females were also friends with males, and the cementing of that first friendship was a new male's key to acceptance. Even if such friendships began when the female was sexually receptive, and most did not, they continued throughout nonmating periods. Female baboons are sexually appealing to males for only a short time; a female was almost always either pregnant or nursing, but even so she could have male friends. Tim Ransom had noticed this during his own studies and had called the friendships "special relationships."

I also noticed that although there were friendships between females and males there were none between adult males. Even friendships between males and females seemed less intense and less lasting than those between females.

Peggy's most important male friend was Sumner, who, in character at least, was the male equivalent of Peggy: older, calm, assured and gentle. Infants flocked around him and females relaxed when he was near. Sumner himself had several female friends; Peggy was preeminent, and Constance was next in line. It was a tight little group—Peggy, Constance, their families and Sumner, who assumed the role of social father, or at least mother surrogate.

Adult males are awesome creatures, but seeing Sumner among his friends told another story. He was attentive to Peggy, adding his protection to her own high rank and thus making her nearly invulnerable. She knew she could rely on him if any of the other males tried to bully her.

Sumner made a fine jungle gym for Pebbles—better than Peggy, because there was more hair to burrow through, longer limbs to climb up and a bigger belly to slide down. He didn't even seem to mind Pebbles's lack of respect for his face: the infant would struggle up Sumner's muzzle to get at the nape of his neck, where the hair was thickest, and repeatedly fall back onto his face. As Pebbles grew, she'd take flying leaps off and onto Sumner, using him as a springboard and not caring where she landed.

As well as being a plaything for Peggy's children, Sumner extended his protection to them. In return, they followed him, sat nearby and groomed him. When he was involved with one of Peggy's or Constance's babies, the females who wanted to fondle the infant would clear it with him first, just as they would have with Peggy: a glance, a present and a lipsmack to the "guardian," then, unless told otherwise, a grunt, glance and touch to the infant. Obviously, Sumner didn't feed Pebbles or his other infant friends, and he didn't carry them around, although on occasion a rescue mission might require his charge to take a brief ride.

Peggy's life was filled with friendly faces. She had no real adversaries, _relationship_ but there were degrees of attachment: foremost was her family, then her _hierarchy_ friends, both male and female, and finally her group mates. Surprisingly, there was also some type of bond that extended to females in neighboring troops, whom she would support against male harassment when the troops met.

"All primates are born social" was a truism I'd learned in my first Washburn course, but it was only through watching Peggy and her family that I really comprehended what it meant. Each truly desired to be with others, and derived great pleasure from achieving this goal. But social tendencies are also nurtured, reinforced and sometimes even actively created by the actions of the baboons themselves. Being close to one another was their life's blood, the basic social activity. Proximity was all-important—it distinguished one troop from another, families from friends and friends from mere acquaintances.

The famous Harlow studies at the University of Wisconsin in the 1960s, which used infant rhesus monkeys, had demonstrated that being close was a baby's most important need. An infant reared in isolation from any social contact became severely depressed—even retarded. When these infants were offered a choice between surrogate mothers, the baby always preferred the soft, cuddly cloth mother to the cold wire dummy, even when the wire mother held the milk bottle and nipple. When the two substitute mothers were side by side, the baby clung pathetically to the cloth mother until hunger could no longer be ignored. Then it went for milk to the wire mother, keeping a foot or toehold on the cloth mother if it could, and returning to it completely as soon as possible. Cloth-mother-reared infants were much less disturbed than those reared totally alone or with a wire mother.

But Tessa, Theadora, Thelma, Pebbles and even Peggy and Thea had much more than a comforting cloth mother. Their real mother offered contact comfort, warmth, protection and food from the earliest days. Contact, being close, also gave babies their family. No one introduced Pebbles to her brothers, sisters or cousins. They were simply the ones closest, always resting, sitting, sleeping and feeding nearby.

Grooming was even better than being close. It seemed to be the most important social tool both inside and outside families. It bridged the social distance, communicated friendly intent and gave pleasure, all at the same time. At first, Peggy groomed Pebbles simply to keep her clean. Then, as she grew older, the reassuring comfort of being groomed could calm her when she was distressed or frightened. Thea, Patrick, Paul, Tessa, Theadora, Thelma—even Constance and Sumner—took turns at grooming Pebbles. Grooming became associated with pleasure, comfort and the contentment that only one baboon can give another. It was grooming that helped create and cement the friendship between Constance and Peggy, Peggy and Sumner, even between Sumner and Pebbles. It was grooming that could soften the blow that might rip the social fabric. Neither Patrick nor Colette seemed to care that one had been the winner and the other the loser in a particularly aggressive row. First tentatively, and then with abandon, they groomed each other. Had they forgotten their angry confrontation or were they making amends?

---

Peggy, her family and friends were changing my image of baboons. Through them I developed a growing feeling about each individual as a unique character, one whose personality was molded by his or her individual history as well as by biological makeup.

Attachments between baboons derive from interaction, from spending time together. The more time, the stronger the bond: family ties are more intense than the bonds of friendship, and these in turn are stronger than the ties of the group as a whole. Even at its weakest, the bond remains tough and enduring. I doubt that a baboon would ask to "be alone," require privacy or solitude: these are all peculiarly human twists to basic primate needs. Competition, dominance, aggression—each plays a role in baboon life. Thea, "the bitch," was particularly aggressive, and even Peggy pushed hard when one of her family was threatened. As the dominant matriline, they didn't hesitate to displace other families when it suited them.

Nonetheless, the predominant theme that ran through Peggy's family and the rest of the troop was a peaceful sociality. Could I reconcile the image I carried home with me each night with what I had expected? Where was "Nature, red in tooth and claw"? Where were the huge, dominant males, bluffing and fighting their way to the top, controlling the lives of their compatriots in the process, and forming the core of the group? Instead, there was Peggy, surrounded by her family at the end of the day, a cluster of baboons at the sleeping cliffs, resting, grooming, playing, simply being social; a big male gently, carefully grooming an infant friend, the infant completely relaxed and trusting. How did attachment and aggression fit together?

# 4. Changes

I was changing. I had come to Africa to do my research project; I had no intention of getting involved with the animals. Both my philosophy and my behavior emphasized my distance. It was always extremely tempting to reach out across the species chasm to touch and communicate with the baboons, but I knew that if I did so, there would be a time when it would be dangerous even simply to observe them. Should they view me as another baboon, one with whom they could exchange touches and greetings, they would also see me as being able to accept their aggression, which I could not do.

Yet slowly and almost imperceptibly I had become deeply involved with the animals. Just being with them created a strong emotional bond. It was nothing like the feelings I would have for a pet; they were not pets. They were friends, in a very unusual sense. Unknowingly, they shared the joys of companionship and the intimate details of their lives with me. I laughed at their antics, delighted in the first steps of a new baby, feared that a youthful male bully would go too far and perhaps injure one of my favorite animals. And without my realizing it at the time, being with them satisfied most of my social needs without placing many demands on me. With baboons, as with humans, friendship is based on reciprocity, but I was in the unique position of being part of a group while remaining outside it, receiving many rewards without paying the price. The baboons touched my heart and mind without touching my body.

In the months that followed my acceptance by the troop, I found myself changing in many different ways. I became the intrepid baboon watcher, going wherever the troop went, taking notes, observing, questioning, thinking. The animals, landscape and mood of Africa were altering my attitudes and expectations in subtle and profound ways.

When I first started watching Pumphouse, the troop and I lived the lazy, luxurious life of the rainy season, when food is plentiful. In the early morning, the animals would relax on the sleeping cliffs, grooming, resting, playing and enjoying the warmth of the sun, often until the embarrassingly late hour of eight or nine, when finally one baboon more energetic than the rest would overcome his inertia and start a move off the cliffs—but only as far as the lush green meadow just below. In more difficult seasons, when food was harder to come by, the baboons left their sleeping cliffs earlier and earlier; sometimes I would arrive at six-thirty and find them already gone.

Now the troop's main exertions—and mine—came at midday: a stroll to water a mile away, where they would drink, nap, groom and play before heading to yet another lush spot. At sundown, between six-thirty and seven, utterly satiated, they clustered near the cliffs for a final bout of socializing before going to sleep. The whole day's traveling usually covered less than two miles. On some days I spent my entire time with the troop, leaving it only at nightfall. On others, I would return to the Red House for lunch with Lynda and Tim, catch up on paperwork and return to the baboons at four in the afternoon, staying with them until they went back up the cliffs. No matter what schedule I followed, it was critical to know exactly where they were sleeping in order to be able to locate them the next day. Even if visitors or unexpected problems disrupted my schedule, I still searched the baboons out every evening so that I could make an early start in the morning.

I don't know where I got the idea that I was required to taste all the foods the baboons ate, but I was convinced my reputation as a field worker would suffer if I didn't. The monkeys enjoyed Cynodon, or star grass, that grew locally, and I, too, appreciated its succulent green shoots. The corms of oniongrass were good, too—nutty-tasting; they resembled little shallots. I also liked Grewia berries, which were sweet but not very meaty. Munching, writing, crunching, gazing, I realized suddenly that I was looking at the Kekopey landscape in an entirely different way—as a baboon, eyeing what was next on the daily menu.

As we moved into the dry season, our life-style—the baboons' and mine—changed dramatically. The land turned brown and I followed, exhausted and amazed, as the animals searched their twenty-square-mile home range for hidden treasures, often sparsely distributed over a wide area. A trek of five miles before noon was not uncommon; sometimes we covered twenty miles a day. But what an advantage the monkeys and I had over other wildlife on Kekopey! Given primate brains and hands, we could harvest foods in many difficult places: fruit from trees stretching over sheer cliff ledges, roots buried deep underground, tiny seeds resisting all but specialized beaks or dexterous hands with opposable thumbs. It seemed to me that the baboons ate everything. Every day I added several new kinds of food to the list, collected samples and gave them provisional unscientific names. I also tasted them.

One of the favorite dry-season foods was Opuntia, prickly-pear cactus, familiar to me from California. The baboons were most ingenious in obtaining this well-defended delicacy. They would maneuver around the large, fierce spikes and pluck off the tender new pads, which were green and soft and had pliable little spines all over them. They would then attack the mature pads by climbing a tortuous path to a good vantage point and tweaking the pad off with thumb and forefinger. This approach saved them from being impaled on their food, but also denied them a firm hold on it. They would drop the pad to the ground, climb down, pick it up carefully with both hands and bite off the spines, exposing the juicy center.

I didn't bother to taste the Opuntia pads. They were used in Mexican cooking, so I was familiar with them. I had not tasted the fruit, however, even though I'd seen it sold on street corners in Israel, cut open to display the persimmon-colored centers and temptingly arranged on beds of cracked ice. Picking up one of the egg-shaped, slightly yellow ripe fruits, I decided to try it later, popped the knobbly object into my shorts pocket and went on writing my notes. A few moments later, I felt a tingly sensation on my thigh. I should have watched the baboons more closely. When they knocked the fruit down, they rubbed it back and forth on the ground, rolling it on all sides. With good reason: at the center of those knobs are clusters of tiny hairlike spines, and it was these that had worked their way through my pocket and into my skin. I took off my shorts, threw away the delectable fruit and rubbed all relevant parts of my clothing and body in the dirt. I was glad the baboons had given me such an easy solution.

However, this did not deter me from continuing to try baboon food. Unfortunately, baboons and humans have different tolerances and preferences; after a few bad experiences, the last of which left me coughing and tearing for fifteen minutes, I decided to bring the new food home and try it when I had an ample supply of water nearby.

Much later, back in California, someone asked me how I really felt about baboon watching. "There's not one single boring moment," I answered, and it was true. At times I may have been physically and mentally exhausted, frustrated, dehydrated and sunburned—but never bored. The most difficult aspect of fieldwork was adhering to my policy of not interacting with the animals. It took tremendous determination not to communicate with my subjects. Of course, to some extent simply being there was communicating: I was occupying a physical space in their social world—and, for all I knew, a social space as well.

Baboons have an elaborate system of communication: their gestures, postures, facial expressions and sounds convey a wide range of emotions. By standing in a certain place and not doing *anything,* I, too, was communicating. Nonetheless, I was drawn further into the social network of the baboons than I had ever intended to go. I realized this one day when I was once again following the drama of the Naomi-Ray friendship. I sat down at baboon level to watch them as they became engrossed in grooming. It made a refreshing and relaxing change; from this level and angle the faces looked very different from the way they appeared when I was standing up. Suddenly I felt a gentle touch on my back, too light to be the rough grass rubbing against my thin cotton T-shirt, and too strong to be an insect crawling inside it. I turned around quickly—and startled little Robin, Naomi's juvenile daughter, who was a few inches behind me. She retreated quickly, and I, just as quickly, figured out what had happened. Robin was very relaxed with me because I always seemed to be around her mother, and she had approached me quietly and tentatively begun to groom the lower part of my back. It was a gesture so intimate that it touched and thrilled me. But I was also upset that I had let it happen, worried that it might permanently change my relation to the troop and ruin my stand against interaction. At the same time it seemed to me incredible that a baboon had trusted me enough to cross the many barriers between us.

After Robin's offer of friendship, I became much more alert to possible overtures from the animals. This was difficult, because the more they accepted my presence, the more they tried to interact with me. At dusk, near the sleeping cliffs, everyone's guard was down and the juveniles especially were eager to establish contact. Adults sat close together, resting or socializing, while the youngsters, exactly like human children just before bedtime, raced around in last-minute boisterousness. They would often approach me as I sat making my final notes for the day, sometimes reaching out and trying to touch my shoes, even grabbing mischievously at the laces.

Once I understood what was going on, I always got out of their way. Their trust and acceptance was a special gift to me, but I stayed constantly alert, hoping to prevent any further breaches—while secretly wanting just the opposite. Most of the time I was able to anticipate situations where the baboons might try to communicate with me, and gradually I learned more about how to avoid or abort these moves by increasing the distance between me and a potential partner. Like humans, baboons have a preferred interaction distance, the normal space between communicating individuals. Sometimes the space is reduced, as in grooming, but there is always some outside limit, so that placing myself beyond this distance was enough.

Once in a while a monkey would become particularly insistent, and when this happened I would employ the usual baboon brush-off technique. One feature of the baboon system of communication is that it is difficult to convey anything to someone who won't look at you. When a baboon prefers not to know what is going on, or would rather stay out of a particular situation, he or she simply turn their backs. It is almost like becoming invisible, and even when threats are involved, it is a very effective way of ignoring attempts at communication.

Large juvenile males could certainly make a nuisance of themselves. About the time they reach the same size as adult females, they try to change their status by pushing around females or youngsters belonging to families which, until then, have been dominant to them. Although deadly serious, these attempts are usually ineffective, especially in their early stages. Females often ignore the threatening juveniles; frustrated but not quite finished, the young male is overcome by the baboon equivalent of ranting and raving. No one will listen to him and finally, exhausted, he gives up, at least temporarily.

One young male decided that I might be easier bait than the high-ranking female he was trying to entice, so when he gave up on her, he started on me. As usual, I moved away. He moved closer; I moved again. Plan A was definitely not working; he tailed me wherever I went. Finally, I switched to Plan B. I refused to look at him or even let him believe that I *could* look at him. He threatened; I shunned. We both knew what this meant, but he was stubborn and refused to give up. We moved around in small concentric circles, the juvenile all the time taking up a new position directly in front of me. Just when I was getting quite dizzy he moved off. I was convinced; sooner or later you could control what was going on by refusing to acknowledge it.

––––––––

Not interacting with the baboons was the most difficult part of my new life as a field observer, but there were unexpected rewards. Attachment, although troublesome, was one; integration was another. Up until this point, my life had always been fragmented. Even the best days in Berkeley were filled with rushing here and there in a futile attempt to accomplish even a quarter of what had to be done every day. My personal and academic lives—each had its own compartment. Even my dreams were boringly the same: of never-ending lists and incomplete lists, of forgetting lists and losing lists. But now there was only one thing I needed to concentrate on: watching baboons, thinking about baboons, writing about baboons, organizing my data on baboons, reading about baboons and, of course, eventually dreaming about baboons. There was another aspect to the integration. What kind of work allows both intellectual and physical challenge? Previously my mental excitement had always been contained within walls; time would have to be set aside for exercise, for unwinding, for burning off the adrenaline generated by the mind at work. Now the two were in step. I would return home each evening mentally and physically exhausted, but wonderfully so.

Unlike most "professional" activities, where months, years or even decades could elapse between starting out and achieving one's goal, each day offered real progress in growth and understanding. Perhaps it was my expectations that had changed; by now I was content to learn one new fact every day rather than groping for earth-shaking conclusions. When your world *is* baboons, your life undergoes a complete reorientation. I had thought that I would miss restaurants, concerts, movies, parties; instead, I began to feel that they were simply poor substitutes for the "real" thing, nature itself. Going to Nairobi, the big city, was a treat. There would be a real restaurant, a movie and friends. I could tolerate one day of it, then would rush back to Kekopey, Gilgil and the Red House like a junkie needing another fix.

I was developing considerable physical stamina. The movements of the baboons during the dry season were far-ranging, and often at a rapid clip. I didn't think about how I was going to keep up with them, I simply did it. The fears Matt had tried so hard to instill never bothered me again. Being with the baboons was, in fact, the best protection I could have had from snakes or any other animal that might cause me worry because sixty pairs of eyes were constantly alert, monitoring and ready for any threat. They saw everything before I did, but they could not save me from the most dangerous animal around: myself.

Falling into pig holes *was* dangerous, not because of the warthogs who lived there, but because it was an easy way to twist an ankle or break a leg. Yet fall into them I did, with disgusting regularity. The more interesting the baboon drama before me, the more careless I became. The first time I fell, I shocked myself as much as the baboons, yelping and ending up in a sprawling heap. I tried—unsuccessfully— not to do it again, since it really frightened the animals.

When one knows a place intimately, it ceases to be frightening. Now I was beginning to be on an intimate footing with Kekopey and I felt more at ease there than in the town of Gilgil itself. People, not wild animals, seemed dangerous. What is unpredictable is disturbing. The more I learned about the Pumphouse Gang, the more comfortable I felt with them and the more uncomfortable I felt anywhere else, including where I rightfully belonged, with people. I have never understood human behavior as well as I came to understand baboon behavior.

Not the least important transformation was the sense of competence and self-reliance that coping with problems in a Third World country fosters. For a decade I drove a car in California and never changed a tire: one phone call and the Automobile Club did it. Here there was no one to call when I drove over some thorns on my way home from baboon watching. Changing a tire is no major accomplishment, but for me it was a triumph. There were curses and tears, and the operation took three times as long as it should have, but I did it, and the next time did it far more efficiently. I even learned some rudimentary mechanics—I had to. If the car stopped dead in a remote spot, I could either walk home for help or sit down and hope someone would come along, so trying to solve the problem myself seemed imperative. I never became an ace mechanic, but many years later, when I was well seasoned and took all this for granted, the oohs and aahs of a friend visiting from California at my prowess in changing a flat, and her unbridled admiration when I cleaned the distributor cap, reminded me just how far I had come.

---

I had arrived in Kenya a contradiction of the stereotype the media loved to portray. Miss Whosits among the Wild Whatsits is in love with Nature. She comes to the wilds to escape Culture, which is Bad, and to commune with Nature, which is Good. Everything is in perfect harmony and so is she, reveling in dusks and dawns, rainstorms and rainbows. In some way, through a combination of reality and fantasy, both carefully nurtured by the media, women field workers had become typecast. I wanted none of it. But being with the baboons led me to fall into the rhythm of wild Africa, a rhythm in which the environment is welcomed rather than avoided. Unwittingly I, too, began to embrace Nature, to shed my cultural baggage and revel in flowers, sunsets, big skies—and, most of all, in the baboons. I noticed this change first in the season of the long rains.

Rain had been anathema to me. It was a nuisance, an interruption, a condition to be got around or through as fast as possible. Rain on Kekopey brought new problems. Difficulty in recording information. Mud. In light rain the baboons simply went about their business, but in downpours they sought whatever shelter was available: a bush, a rock or even the sleeping cliffs. The long rains began six months after my arrival. As their name implies, these were the serious rains, lasting from April through June and depositing huge amounts of precipitation. But by now I had undergone a sea change, or at least a rain change. I welcomed the moisture after the long dry period, appreciating its effect on the plants and animals. I watched the storm front moving across the landscape and was thrilled. I stopped fighting the rain and let it refresh me, just as it did everything around me. And I too fell into that romantic image of nature, the media's packaged view. Nature was a pristine world; everything was in perfect harmony except when humans interfered. Certainly what I was discovering about the baboons was reinforcing this view rather than contradicting it.

But there was another side to nature, one that the media usually ignore: for every birth there is a death, and death itself is rarely pleasant. Events forced me to see it as part of the natural cycle and to ground myself in the real nature rather than in a romanticized ideal of it.

The first death I witnessed was that of Lisa. She had fallen from a tree, and was so stunned that she didn't present to the adult male who had chased her along the fragile branch from which she had fallen. He nipped her when he saw her, which was usual, but she was so dulled by the fall that instead of holding still and allowing him to attack her well-protected behind, she turned toward him, and his canines sank into her side, penetrating her intestinal wall.

Lisa was able to keep up with the troop for several days, but gradually she began to lag farther and farther behind until she was too weak to move. I stayed with her, taping my notes about her final hours. As I listen to that tape today, I can hear the quiver in my voice and the pauses where I had to switched off the recorder.

Lisa and Sickle were best friends, like Peggy and Constance. What would Sickle do that evening when Lisa didn't arrive at the sleeping cliffs? The troop moved from their drinking spot under the big acacia tree, but her friend lingered behind, looking toward the departing troop. Sickle did not leave her lookout post until darkness fell and the rest of the baboons were safely lodged in the cliffs for the night.

The troop's reaction to the death of Quentin confused me. I couldn't tell whether baboons, like chimpanzees, recognized death. Quentin was bitten by a snake and died at a watering spot the troop used frequently, and to which they returned the following day. As they came down to death drink, each monkey seemed to treat Quentin as if he were some strange new object in the environment, bending down to sniff and investigate, then moving on. Queenie, Quentin's mother, and her family arrived next. At first they approached the body in a similar way, but stayed around it, circling, sniffing and clearly disturbed. Did they recognize Quentin? What did they feel? By this time the troop had moved so far away that the family had to race to catch up.

I remained confused after seeing the troop's reaction to the death of a baby. While Peggy had her hands full with her new infant, Pebbles, Harriet—a middle-ranking female—gave birth to her first baby, Hal. Harriet had had plenty of experience handling other infants in the troop, so it surprised me to find that she lacked the calm assurance and maternal skills I would have expected.

On the first day of Hal's life, Harriet held him upside down, with his head between her legs and his bottom where his head should have been. Naturally, Hal was distressed and so was everyone else. His cries finally roused Harriet, and she righted him, but her awkwardness persisted. She had trouble holding Hal with one hand and walking on her other three limbs, and being the rest of the troop's center of attention made things worse. Harriet seemed unable to relax into motherhood.

Then there was a crisis. Naomi came to pay her respects. Since Naomi was of lower rank than Harriet, the latter had no qualms about letting her handle the baby, but after a while Naomi didn't give Hal back. Baboons are quite restrictive with their infants, unlike langurs and Colobus monkeys, who pass a baby around from the day it is born, seeming only too happy to get rid of it for a while. A baboon mother gives up her youngster only to someone she trusts completely: a friend, family member or an individual she is able to intimidate. Naomi, Harriet probably figured, fell into the last category.

But Naomi would not be intimidated by Harriet's threats. She simply ran a little farther off, Hal clinging to her tightly. This continued for over an hour, and no one came to Harriet's aid—in itself an unusual event. Finally Harriet gave up, but didn't go away. She followed Naomi and Hal tenaciously for several days, moving a little closer every time Hal squealed, which he did frequently. As the days wore on, Hal, who had known his mother only briefly, gradually relaxed on Naomi's belly. Because she was in the late stages of pregnancy herself, Hal's sucking on her nipple produced a milk flow, and it looked for a while as if the adoption might succeed, especially since Harriet appeared to be losing interest; occasionally she would stop following Naomi and go to sit with her older sister or with Rad, her closest male friend. But no matter how far away she was, when Hal made any noise Harriet would rush back again and try once more to retrieve her baby.

I came to learn that this whole situation was extremely unusual; females simply did not kidnap one another's babies. In fact it was so unusual that no one in the troop knew what to do about it. On the morning of his tenth day "in captivity" Hal died. Why? How? These were questions to which I would never learn the answers. He didn't die of starvation; if Naomi's milk had been inadequate, he would have died long before. Was there some congenital abnormality? If so, could it explain Harriet's awkwardness with Hal in his first days, or even the whole series of bizarre events?

Naomi clutched Hal's lifeless body to her, and his arms and legs dragged on the ground as she moved. She continued to carry, groom and tend Hal as if he were alive, occasionally placing him on the ground so that she could feed more easily, but scooping him up as soon as anyone approached. Harriet shadowed Naomi, showing more interest in Hal than she had for a week. This interest lasted only a day. That evening, Naomi simply put Hal down and didn't pick him up again. No one else was around, not even Harriet, and that was the end.

Dead infants are often carried for days; at first, the body is held close and groomed, but as it begins to decompose and attract flies, the mother places her infant on the ground to make her own feeding easier. This behavior—as I had seen with Naomi and Hal—continues until the mother abandons her nearly unrecognizable infant.

To a baboon, death means the loss of social identity. A corpse was either a nonentity—as Quentin's had been to the troop—or an enigma, as Quentin's was to his family or a newly dead baby to its mother. It was Quentin yet not Quentin—my baby but not my baby—until it had changed so much and was so inert that any recognition or attachment was no longer possible.

Injuries posed the same dilemma for the troop. When Clifford, one of the juveniles, injured his leg, forcing him to run rather than walk, all his age mates ganged up on him, taking advantage of his inability to fight back. His mother, Constance, put an end to such behavior, but only temporarily, and Clifford continued to find himself the butt of the larger males' games. This happened again and again, with adult animals as well. Bo, an adult male, had a similar leg problem; females and adolescents fled screaming from him, and the other adult males made him the scapegoat in their aggressive encounters. It was easy to see why the females reacted as they did: before his injury, Bo had made a dash for them only when he was furious, so it had always been a good idea to get out of his way. From the males' point of view, this once-feared animal was now obviously unable to fight back, so old scores could be settled and new ones racked up.

No matter how peacefully sociable baboons are, no matter how far removed they are from our usual image of aggression, they are not angels. They seem not to forget some old disagreements. The fact that communication between them is so complicated is important to an understanding of their reaction to a troop's sick and disabled members. When illness or injury forces a baboon to significantly alter gestures, postures and behavior, something unpremeditated gets communicated, and the other baboons respond to the altered message.

These deaths and injuries among the baboons aroused feelings of sadness and pity—human emotions—in me. It was taking me a long

*injury =>*
*altered*
*communication*
*=> altered*
*social*
*responses*

time to accept the natural rhythms of the Pumphouse Gang, where death was a part of life and not simply ritually absorbed.

Perhaps the most difficult aspect of my confrontation with death was during predation by the baboons. For a long time, anthropologists had believed that of all the primates, only humans were meat eaters; certainly, they agreed, only humans were *hunters*. But Jane Goodall's documentation of hunting and meat eating by the Gombe Stream chimpanzees forced us to modify this concept by showing us cooperative chimp hunters at work. I was to find similar behaviors within Pumphouse.

My first encounter with predatory behavior came when I was following Peggy, trying to keep up with her while the troop was moving at a fast walk and simultaneously recording her actions in a complicated code. She put on speed. Was she fed up with my shadowing? Had I offended her? But she wasn't glancing back at me as she would have done if I'd been the reason she was moving so quickly. She was making straight for two males who were sitting unusually close to each other.

Peggy arrived long before I did; the only way I could keep up with even the smallest monkeys when they were walking fast was to run, and the baboons didn't like that. So I had to keep to a quick walk, without trotting.

A grisly scene greeted me. Sumner sat in the middle, a squirming baby Thomson's gazelle in his mouth. He had it by the neck, its head near his mouth, its legs flapping. The next minute, the tommy was pinned to the ground, front and rear legs held down by strong baboon arms. Sumner was eating from the soft underbelly while the baby was still alive. It did not die until five minutes later; there was no quick killing bite. Eating proceeded quietly.

I was horrorstruck. I knew from Bob Harding's observations that the Pumphouse Gang was among the few predatory nonhuman primate groups that had been studied in the wild, yet I was unprepared for what this actually meant. It didn't help that the prey was a young antelope a little like Walt Disney's Bambi. As long as the fawn looked like a fawn, I reacted profoundly, but as Sumner gradually worked his way from the belly to the ribs to the neck and the limbs, turning the carcass inside out in the process, the baby slowly became a piece of meat and I could treat it with more detachment.

By now a crowd had gathered. When Peggy arrived, Carl, an older male somewhat resembling Sumner, had already staked his claim by sitting very close to the other male. This made Sumner nervous; glancing at Carl, he slowly but definitely turned his back so that he could eat the tommy without having to look at Carl. But Peggy marched right in, positioning herself in front of them and forming the third corner of the triangle.

Ten minutes passed. The longer Carl remained, the more nervous Sumner got. He glanced frequently at the other male and started to eat faster; it was as if he was trying to eat as much as he could before letting Carl have the carcass.

Peggy sat by calmly, not exactly staring at the carcass but clearly intent on Sumner. Occasionally he moved a few feet, leaving behind bits and pieces on which Carl swooped like a vulture, stuffing what he could into his mouth and then continuing to tail Sumner closely. Whether it was intended or was accidental, these moves gave Sumner a brief respite from Carl's overbearing presence.

Finally Peggy got her turn—not that Carl had left much. A close examination of each feeding place usually turned up a fragment of bone, a small piece of meat, a bit of ligament. After retrieving what she could, Peggy resumed her part of the triangle. Then Thea arrived, more nervous than Peggy had been, but clearly interested. Patrick and Tessa also put in an appearance, but decided that chasing each other down the hill was more interesting. The audience was increased by the arrival of several juvenile males. Sitting off in the distance but with both eyes riveted on the carcass was Frieda, a middle-aged, middle-ranking female with a peculiarly short tail.

Sumner left the carcass without any warning, and Carl moved to the tommy and began to eat. Sumner didn't even glance back. With his cheek pouches bulging, he strutted, pulling himself up to his full height, to the shade of a nearby tree, where he leisurely consumed what remained in his mouth.

There didn't seem much left to eat. The whole animal looked as if it had been peeled; the pelt was turned in on itself down to the hoofs, exposing the long, thin bones and delicate muscles of the legs. All the other bones, muscles and intestines had been eaten. The still-intact head was the only reminder of the creature's original beauty. Carl started in on this and my throat tightened again. First the ears, then the eyes and lower jaw were quickly dispatched. The beauty became an unrecognizable blob. I heard a crunch; Carl had lodged the skull between his molars and pressed down. When the thin braincase cracked, Carl devoured its contents, then nonchalantly discarded what remained and went off to find his own shade tree.

A scramble ensued. The juveniles and various females made a dash for the carcass, but Peggy quickly claimed ownership, being the closest and most dominant female present. I couldn't imagine what there was left to eat, but Peggy proceeded systematically to crack the long bones and eat the marrow. She reexamined the pelt and all the bones, cleaning them of any small pieces of meat before she, too, relinquished her prize. The tattered heap of fur and hoofs was then investigated by each and every attendant in order of their dominance ranking; they all found something, but the period of carcass ownership decreased dramatically, until the last small juvenile gave it the merest once-over before rushing away to join the rest of the troop.

Now it was my turn. The tommy looked as if it had been professionally skinned, with a slit down the middle and along the edges, but these professionals had been careless: the hoofs were still attached. I couldn't see anything left worth eating. A pile of splintered bones remained at the spot where Peggy had been feeding, and the carcass had been abandoned not far from there.

Nothing remained of the vibrant little tommy that had been alive an hour before. In my fascination with the events, I had briefly forgotten my original horror. Was there a right and a wrong? If so, who was right? Were the baboons wrong to eat the tommy? Worse yet, to have eaten it alive? Did they *have* to do it? Wasn't there enough grass, flowers and fruit to satisfy them?

The second time was no easier. Perhaps if the baboons had killed the tommy before they began to eat it, I might not have reacted so strongly. But my feelings resembled that of the mother gazelle who, as long as the baby was still moving, sometimes even bleating, stood around, at times charging the predator. Once all activity ended and the baby was a lifeless body, the mother would run off and join her herd, this chapter in her life ended.

The third time was easier, and the fourth and the fifth. By the sixth killing I had become objective, a distant observer of baboons, unmoved by the plight of their prey. My mind was fully occupied with the details of the kill: how it was captured, who ate it, who got a share and who tried for one and failed. Even those first terrible minutes became easier to watch.

---

Death. Like most city dwellers, I had been insulated from it. My meat came neatly packaged, bearing no resemblance to the animal it once was. Intellectually I knew all creatures had to die, and was familiar with the theories that had been developed to explain predator-prey interaction, life histories and population dynamics. But I certainly did not look forward to dying myself. Would any creature? Is death without awareness less painful to the living? Was a gazelle or a baboon truly unaware of what was happening?

Seeing death and birth together, observing the sequences that connected both to a life cycle, made it part of the same whole. Death did not intrude like some avenging angel, it was only a process, neither good nor bad.

Growing up "civilized," not wishing to embrace nature, having no value judgments, I had never taken sides. Yet the baboons, by turning my world on its head, had changed me from an aloof intellectual to a naïve romantic. When the Maasai boys brought me an orphaned baby zebra, I didn't understand why Tobina Cole advised me to kill it. I hadn't realized how kindly she had meant it until Chloe, as I named the fuzzy-nosed creature, died after being marched through the blazing, sun-baked streets of Nairobi in a parade—a parade in which the Animal Orphanage was participating.

The baboons led me still further. They saved me from falling into the trap of dividing the world into false dichotomies: into nature and culture, animal and human, good and evil. When the baboons came to the Red House, they made no distinction between trees as playthings and the house as a giant jungle gym, between the empty bottle they found in the garbage pit and the large, interesting twig they also used as a "toy," between the water in the drainage ditch and that in the rain puddle, between the natural and the "artificial."

Nature, I was learning, wasn't a violent struggle—everyone against everything—and yet it wasn't free of unpleasantness, either. What *was* natural in this animal world, and where did humans fit in?

# 5. Issues

By January 1974 I was thriving, emotionally and physically. Every day brought its own challenges and discoveries—which was, I suspected, going to be the problem. Each day I spent with the troop it became clearer: sooner or later I would have to confront the fact that I was seeing things I was not supposed to be seeing, finding patterns that were not supposed to exist. Worse still, I was *not* finding patterns that everyone else said did exist. The intellectual framework I had brought with me from Berkeley had become unrecognizable.

With Washburn as my guide, I had been introduced to the Primates, the biological Order to which humans belonged. Primates appeared sixty million years ago as tiny, insignificant creatures very similar to tree shrews. They were among the earliest of the primitive mammals who filled the vacuum left by the disappearance of the dinosaurs. I had learned the differences and similarities between ourselves and the early prosimians, about how a new way of life—living in the trees of the tropical forest instead of on the ground—transformed anatomy and behavior and set the general primate pattern that we humans, living millions of years later, still share. Instead of relying on the sense of smell, primates depended on vision to orient themselves in their world, a three-dimensional world requiring special acuity. Reaching for a branch high up in the tree canopy could be dangerous if you missed. Depth perception was critical, as was color vision.

I was surprised to learn how few animals saw the world in either of these ways. In order to perceive depth, eyes have to face forward so that visual fields overlap. I was disappointed to discover that the bull in a bullring saw only the motion of the red cape and not its color; to him it was gray, and the color excited only the spectators.

The earliest prosimian fossils so far discovered have long snouts and eyes at the sides of their heads. Clearly they relied on smell and vision, even though they could not see stereoscopically. It was not long before new creatures appeared; these had large round eyes that faced forward and nearly flat faces. That they could see colors is a guess, but their reliance on vision rather than smell and their possession of depth perception are both reasonable deductions.

The rest of the body underwent major changes, too. How do you get around up in the trees? Squirrels dig in with their claws and pull themselves up. Small primates climbed by grasping. To do this, they needed specially formed hands. The primate hand is a remarkable invention, and one that helped forge a new relationship between the animal and its environment. The hand has mobile digits that can move independently of one another. The thumb is distinct, and opposable to the fingers. Instead of grappling hooks—claws—there are flat nails that protect the sensitive tips of the fingers without getting in the way. The palms and fingertips not only have pads for grasping, but their generous supply of nerves allows them to feel as well. Hands and feet are very similar, but the former are clearly the more dexterous.

As areas related to vision and manipulation increased in size, as more sensory information was processed and integrated and as links needed to be made between sensation and action, the brain itself became more complex. The hallmark of a primate is its hands, face and brain: no other type of creature on earth possesses this unique combination. Together, these formed the basis of the primate pattern, which heralded not just a new capacity for exploring and dismantling the world, but a totally new *Umwelt*—a new way of perceiving everything: you, me and it.

A change in behavior and anatomy is a bit like a chicken-and-egg story. I marveled that Washburn could interpret so much from the handful of bones he showed us. He constructed a sequence based on the dates that accompanied each creature. In isolation each bit told little, but taken together it was an exciting story, one that captured my imagination. Prosimians, monkeys, lesser apes, the great apes, early hominids, true humans and modern humans—every step along the evolutionary sequence showed that there were important lessons to be learned about human behavior.

The story Washburn unfolded always returned to the same provocative question: Why are we the way we are? His answer: Humans were an accident. Although we began with a certain primate resourcefulness, our final outcome is not preordained. If anything, we will probably fail: over 95 percent of all animal species have become extinct. Why should we be any different?

If we were to succeed as a species, it would be because our behavior would allow us to adapt to changes in our world. That was our ace in the hole. Just as we inherited primate anatomy, with all its advantages, so we inherited certain advantages of primate behavior. But in order to understand what this meant, we would have to know a great deal more about primate behavior.

Some aspects of behavior could be guessed at by deciphering anatomical and paleontological clues. Today's primates help us with this task by providing insights about behavior and how behavior and anatomy are interrelated. Living primates are not relics of an ancient time. The prosimians alive today have continued to change during the sixty million years of their evolutionary history. So have monkeys and apes, but over a shorter span of time. To say that we are descended from the apes, as Lord Wilberforce did in a taunt to the proponents of evolution over a century ago (and he was referring to the living apes) is to misunderstand the concept of evolution. The study of living primates can contribute an understanding of the primate pattern; of what is shared by all species; of what has changed in the different species and of what is unique to any one species.

Equally important is the fact that we can understand anatomy only by observing how living creatures move through their environment, how they solve life's problems. How does a body built like that of a tarsier—a nocturnal prosimian with huge eyes and elongated legs—actually work? Some aspects of tarsier anatomy are remarkably similar to human anatomy, although the two species are only distantly related among the primates. What can this mean? Gorillas and gibbons have similar shoulder blades. We know that gibbons can swing lithely through the forest from the underside of one branch to another. Could a creature as large as a gorilla do the same thing? Anatomically it *should* be possible, but in reality what do the animals do? What are their behavioral and anatomical constraints? The questions are endless, some more critical to our interpretations of our past, some less.

Certain crucial behaviors don't fossilize. Where would you look for the fossil of mothering, of sexual encounters, of Ray and Peggy? Anatomy provides hints of what is possible and what is not, but anything further is wild speculation. In these areas our only windows into the past are the living relatives of those fossil ancestors.

The study of primate behavior is a relatively new phenomenon; a few field studies had been done before World War II, most notably by C. R. Carpenter, a psychologist who worked on gibbons, howler monkeys and spider monkeys in the 1930s. The 1950s saw a spate of studies, primarily on monkeys in Africa and Asia. George Schaller worked with gorillas, one of the great apes of Africa, and a variety of abortive studies on Asian apes were undertaken. The Japanese were studying the monkeys native to their own country and were sending their scientists to investigate primates in other parts of the world. But Sherwood Washburn and his students were the first to become actively engaged in these field projects with a view to understanding human evolution.

Expectations in those early days were very different from what they came to be later. A season's work seemed reasonable, although one year's coverage was a desirable goal. Most of what was known about wild primates before then was simply anecdotal. In the new studies, the choice of species to investigate was dictated as much by opportunity and convenience as by any set of priorities, since it was felt that a great deal could be learned from *any* species. True, some species seemed more relevant to questions of human evolution than others. Our unique human adaptation developed when our ancestors came out of the forest and began living on the African savannah. This has happened only a few times during primate evolution: aside from the hominids—our immediate ancestors—baboons and patas monkeys are the only primate inhabitants of the open veld. Vervets can be found in the narrow strips of riverine forest that penetrate the bushland and open savannah, but they don't actually live in the open away from trees. Baboons are particularly interesting because they forage out in large groups, facing the dangers of savannah life through group action, much as we imagined our ancestors did, unlike patas monkeys, who live in small groups and survive by stealth.

Although baboons are biologically more distant from humans than chimpanzees, their life-style and their ecological setting, thought to be of critical importance in determining behavior, made baboon models central to reconstructions of the earliest stage in human evolution. Several baboon studies combined the expertise of observers trained in anthropology, psychology and zoology. During this formative period in the development of behavioral reconstructions of human evolution, baboons were *the* model, as well as the most studied nonhuman primate.

Between the 1950s and early 1960s and the time I set out for Kenya, important changes occurred in our view of nonhuman primates, including baboons. The more we learned about wild primates, the more variation there seemed to be, not just *between* species but *within* species. It became increasingly difficult to trust all the grand generalizing.

When Washburn and his student DeVore wrote their early papers on baboons and human evolution, they were confident that studying baboons told them something important and specific about human evolution. As more and more variations were documented, it became harder than ever to feel justified in making such comparisons. If the behavior of baboons in one place didn't accurately predict the behavior of baboons living somewhere else, how would we be able to use such behavior patterns as models for comparatively distantly related humans? A number of first-class scientists deserted the field, deciding to study humans rather than animals in an effort to understand human behavior.

At the same time, the chimpanzees' star was rising. Anatomically, chimps and humans share a number of characteristics distinct from monkeys or prosimians, which can be seen in the organization of the shoulders, ribs, pelvis and, to some degree, the hands and feet. And chimpanzee brains, while small by human standards, are nonetheless several degrees above monkey brains in both size and organization.

Jane Goodall's study of wild chimpanzees at the Gombe Stream in Tanzania, begun in 1960, revealed many new behaviors closely resembling those of humans. Although primarily fruit eaters, chimpanzees did occasionally hunt and eat prey. They used tools and even shared food, two characteristics previously thought to be uniquely human. In addition, genetic studies, especially sequencing DNA, demonstrated an even closer biological affinity between chimpanzees and humans. Some scientists estimated that chimps and people shared 96 percent of their genetic material, while other scientists raised the percentage to a startling 99 percent. Had this been true of any other pair of animals, the two would have been classified as sibling species, but humans like their uniqueness. The old hierarchy of beings evident in the Bible, in Greek science and even in the Renaissance is alive and well in modern science. The Supreme Being comes first, then humans, then the nonhuman animals, ordered by virtue of their ability to act like humans.

Chimps definitely posed a problem. Each time chimpanzees demonstrated a humanlike ability, the definition of what was human changed. Man the hunter became man the toolmaker became man the food sharer became man the user of language. Even this last definition is under attack now that a variety of apes can converse with their human observers in American Sign Language, or by symbols on a computer or board.

Although chimpanzees replaced baboons as the model for reconstructing the evolution of human behavior, baboon studies and the baboon model have had a surprisingly lasting influence. I often wondered why this was so, and only later gained some insight.

Baboons are ubiquitous in Africa, from the arid regions of Ethiopia all the way down to the tip of the continent. If numbers and distribution count, they are second only to humans in their success as a primate.*

It seemed that baboons could adapt their behavior to many different kinds of environment without having to change much of their basic anatomy. They are definitely a primate success story. By comparison, chimpanzees are on the verge of extinction, so where has all their near humanness gotten them?

In the heyday of the baboon model, ideas about primates reflected baboon studies to a disproportionate extent. When you read about "the primate male," "the primate female" or "the primate group," you were for the most part reading about baboons and about savannah baboon society. Other nonhuman primates were known to live in a variety of

*True, Asia has its baboon equivalent—the macaques, sometimes called the baboons of the East. Although they live in the same variety of environments and climates, there are thirteen different species containing more than forty-five subspecies, each uniquely adapted to its own particular environment. This compares to only two, three or five baboon species, depending on which taxonomist you listen to: savannah baboons (including olive, yellow and chacma), hamadryas—and the Guinea baboons. There are three other monkeys—mandrills, drills and geladas—which closely resemble baboons in many respects.

styles, such as mated pairs (like some prosimians and even the advanced lesser apes) or single male groups, where several females shared the same male as mate and protector, but it was the multi-male group as represented by baboons that was taken as the norm of primate social organization. Several unrelated adult males lived with many more females and their young in a cohesive and well-organized group. It traveled as a unit, feeding, resting, moving, sleeping together. The group was a primate's greatest asset, a resource for assistance in all aspects of life: foraging, socialization, protection. Baboons illustrated just how this would work.

But the lasting appeal of baboons has more to it than that. The books that continued to pay homage to baboons had a message: our human origins hark back to a male-dominated society with a clear division of labor, one in which males hold all the power and females acquire status only through their association with a "dominant" male. In this society, males vie with one another, using force to attain dominance and claim their rightful spoils. What expectations should we have of modern humans if a killer ape (actually a killer baboon) still lurked within us, as Robert Ardrey has suggested? What if only primate males have political prowess? Could there ever be a woman president; should we even allow it? What if females are evolutionarily so constructed as to be able to do nothing except rear babies? These ideas influenced an entire generation in the fifties and sixties.

The original baboon studies by Washburn, DeVore and an English psychologist, Ronald Hall, created a neatly constructed picture of baboon society. Yet even in 1972, when I was starting out, the fact that this idea should still be considered of such importance, especially in the face of all the new primate evidence, distressed me. As I traced the history of ideas about baboons, I found that the picture changed and became simplified the more often it was retold by the scientific and lay community. There was no mistaking its compelling message: males [Old Theory about Baboon ranking] were the building blocks and the cement of the group. They were the focus and the power. They were the structure and the stability, the essence and the most valuable part.

Male baboon bodies were fighting machines, with powerful muscles, thick mantles of hair and razor-sharp canines. With such equipment, males competed with one another for whatever good things there were: [competition] food, females, a place to sit. Although the most effective way to compete was through aggression, males did not seem to fight *all* the time. In fact, [(aggression)] once they *had* fought, they achieved a certain rank; thereafter, lower-ranking males gave way without protest, while those of higher rank confidently strode up to their rewards. This male-dominance hierarchy, arrived at as a result of aggressive contests and maintained through threat and bluff, was what gave the group its social structure.

*females*

What were the females supposed to be doing? Their lives were meant to revolve around babies: bearing them, nurturing them, socializing them into proper adults. There seemed to be a certain number of rather subtle relationships among the females, but they were always subordinate to the males. A temporary sexual liaison with a male could be

*subordination* useful, for his company could result in greater protection or more or better food, but this rise in status was short-lived.

All eyes were focused on the males. Females and youngsters jockeyed for positions near a dominant male and vied for the right to groom him. It was to their advantage to do so: the males watched out both for external dangers and internal strife; they were the guardians and the protectors.

*concentric*
*social order*

This social order and the differences in sex roles were constantly visible, the early studies noted, in the way the group moved in its environment. A baboon troop on the move resembled a series of concentric circles: in the center were the dominant males and the females with young babies; surrounding them were the rest of the females, older youngsters and lower-ranking males; finally, on the periphery, were the least important young and adolescent males. Should a predator surprise the troop, it would certainly snatch up one of these peripheral and clearly expendable members, while if the danger was detected early enough, as few as three full-grown males, canines flashing, bodies nearly doubled in size by their erect mantles of hair, presented a sight formidable enough to chase away any intruder.

A few other studies painted a different picture. Thelma Rowell had observed the same species of baboons as Washburn and DeVore, but hers lived in a forest. In such an environment, males were the first to flee from danger to the safety of the trees, leaving the females and youngsters to fend for themselves. Dominance between males seemed less clear-cut, less powerful and pervasive.

Tim Ransom's study of baboons at the Gombe Stream Reserve exposed a wealth of complexity in baboon social relationships, including observations that interactions between females were not as subtle as they had first seemed to be. His research supported the conclusion of studies on the free-ranging macaques of Cayo Santiago and the Japanese macaques, both of which demonstrated that females had dominance hierarchies.

What about baboons? The transformation of baboons in the early baboon studies to the simple picture of male dominion, and the tenacity of that picture, whether conscious or unconscious, dovetailed nicely with ideas in Western culture of how the world should be. Modern society should contain a "natural" division of male and female roles, a division into male political power and female domesticity, the interpretations seemed to say.

I had my doubts about this picture of baboon society from the beginning. I knew about Rowell's study, and about the macaques, about Tim's findings. "Go out and watch baboons," Washburn had told me. I was doing just that, and the more I watched the Pumphouse Gang, the less my observations jibed with the baboon "model." At first I had assumed the discrepancies were due to my inexperience and general ineptitude, but time gave me both experience and confidence. These baboons were not playing by the rules.

First of all, it was apparent that males were only temporary members, transients, integral parts of their society for just a brief period. There are no all-male bands outside the group, nor are males particularly *no all male gp.* friendly with one another once they reside together in the same troop. *no ♂♂ friendships*

Since males come and go, they couldn't be expected to provide the stable core of the group. Upon closer examination, it was clear that Pumphouse males certainly didn't. They couldn't. Much of their time was spent in working out their own relationships and trying to achieve some degree of stability among themselves. This did not appear to be easy.

Furthermore, I tried hard to see the neat dominance hierarchy that males were supposed to have, but it was almost impossible to line up the males or to say which male in any pair was really the dominant one. Ray and Big Sam were a good example. One day Ray would come out ahead; the next, sometimes even within the hour, Big Sam would be the winner. Undaunted, Ray would try again until he won. Stubbornly, Big Sam would refuse to let the issue drop.

True, the situation between all the males in the troop was not always chaotic. Carl and Sumner seemed to have everything worked out between themselves fairly well, but as relationships between the other males changed, the balance of the entire troop would be upset. Even if Carl and Sumner didn't reverse their ranks each day, when Ray decided to harass Sumner instead of Big Sam, Carl felt the effects.

It was all very confusing. I could detect no stable dominance relationships among males, and certainly no linear dominance hierarchy. As if that wasn't bad enough, what ranking I could find predicted little or nothing of importance. One of the main keys to an understanding of the whole male-dominance argument has been that males fight for rank because once they achieve it they are thought to have a better chance at getting important resources. In fact, high-ranking males were believed to monopolize *the* most important resource—sexually receptive females. To my surprise, although certain aggressive males might have the advantage in Pumphouse at times, it seemed to be the loser who got the important rewards like receptive females, specially prized food, the majority of the grooming.

The first time I saw it happen with an estrous female, I couldn't believe my eyes. Ubiquitous Ray was harassing Big Sam for all he was worth. This was nothing unusual, except that this time Big Sam was in consort, following and monopolizing a receptive female. I had mastered the description of the scene, using my ethogram of behaviors to record the sequence, the give-and-take of aggressive signals between males. I had told myself smugly that *I* knew what was going to happen next: Ray was winning, and would claim his rightful prize, the female.

He did win, but when Big Sam rushed off to avoid any further confrontation, Ray followed, tailing close behind him for the next hour and leaving the female to her own devices. Before I knew what was happening, Rad, who had been sitting on the sidelines keeping track of all this with what appeared to be calm disinterest, raced up and claimed the female, beginning a new consortship. There were no objections.

Why fight and not take the prize? Why fight at all? I was sure I must have missed some vital aspect of the exchange between Ray and Big Sam, and that what I thought had happened, hadn't. But the pattern recurred: males acted aggressive around some prized resource, and then both winner and loser walked off. Further, I couldn't identify the dominant male in the troop, and no male acted the part. So what was the good of dominance? Why would a male fight and risk severe injury for something that seemed to benefit him so little? And there was a complication within the complication: although the males were aggressive toward one another, they were less aggressive than I had expected. There was no linear or stable dominance among the males; they were much less aggressive than they were supposed to be, and—most confusing of all—aggression between them didn't seem related to who got the rewards they prized. The males were consistent in just one thing: they were not doing what they were supposed to be doing.

That was the bad news. The good news was that the females were much easier to understand, and their role made a lot more sense than before. They and their offspring were the stable core of the group. They had to be; they were the only permanent members. Peggy lived with the Pumphouse Gang the entire thirty-odd years of her life; any son of Peggy's stayed with the troop until he was adolescent, then wandered off to another group. Once there, his stay might be as short as a month or, as I was to find later, for as long as ten years. But ten years was less than a third of Peggy's entire life. No matter how you viewed it, males were only temporary members of a group. The females were permanent.

*permanence => stability*

Moreover, females were predictable. Their first allegiance was to their family. Families themselves had ranking principles; Peggy's family was a good example of this. Mother was on top, the children below in reverse order of age. All the adult females had their own dominance rank vis-à-vis one another. The linear hierarchy I had found to be missing among the males was alive and well among the females, who originally had been assumed to have only "subtle" relationships. Of course there were many "subtleties" that I was still to discover, but knowing a female's rank and her family helped me to predict how she would behave in the vast majority of interactions she had with other families.

*predictability*

The female hierarchy seemed very stable. When a female was in consort with an adult male she might gain some extra immunity from attack by higher-ranking females, but her own rank did not really change: it was more like having a brief vacation from the normal daily routine, nothing permanent or serious. Female roles were more than simply that of baby makers. Matriarchs were not only the protectors of their children, warding off bullies from within the troop and dangers from outside it; they were also the policers, maintaining peace and order within the family and between families. They were the primary focus of family attention, and directly and indirectly wielded a great deal of influence.

*} matriarchy*

Males were not exactly irrelevant, but they certainly were not the prime movers described in the earlier reports. Their sheer size and strength made them dominant to all the females and youngsters, but their sphere of influence was actually much narrower than one would have predicted.

*complementarity*

Who, then, was the "leader," the dominant individual who determined where the troop went and played the main role in its defense and order keeping? No one, male or female, fitted this description. Males and females were not equals; they played complementary roles, one no better than the other. Because females spend their entire life in the same troop, they know their home range intimately, while, as Thelma Rowell has suggested, males know areas outside it that might contain resources critical to survival during difficult periods such as severe droughts. The males benefit from the females' knowledge and leadership on some occasions; on others, they take the lead.

*females - family*
*males - surviv.*

The Pumphouse Gang certainly fitted this description. Ray learned the basics—where to go for food, drink and sleeping sites—by first observing the troop from outside, and then by becoming one of its members. He was never in the lead in troop progressions, except perhaps when he tried to get in front of the group *after* it had made up its mind to shift locations. Yet many months later, when dry-season shortages were making themselves felt and he had made a few female friends, Ray tried hard to influence the females and their children to follow him beyond the edge of their familiar world. He moved out determinedly, glancing back at Naomi and the others, signaling to them to follow. And follow they did, at least until they reached the boundary of the Pumphouse Gang's home range, where they seemed to be held back by some kind of invisible fence. Ray managed to walk through it, but seeing his army lagging behind, he stopped and indicated that they should continue. They refused, and it took him five minutes of come-hither looks, grunts and glances, not to mention several false starts, before he managed to move them beyond the imaginary barrier. Once they got going, they stayed close to Ray as he marched them to a food-rich grove of acacias. Hence Pumphouse's home range was enlarged, at least during this period.

I was not finding a simple reversal of roles in Pumphouse; females were not politically powerful and central while males were powerless and peripheral. The picture was much more complicated and interesting. Females did not rely on males to protect their infants, but those babies were given extra protection *because* there were males in the troop. Most internal disputes were taken care of within the framework of the female hierarchy and family system; if a gross injustice seemed about to occur, or when disputes threatened to drag on unresolved, a male would permit himself to become involved, particularly if he was a friend of one of the protagonists. Both males and females had power, *shared* and both sexes exercised it. Both males and females were involved in shared caretaking, since males frequently became almost surrogate mothers to their infant friends. Both sexes were political, and both sexes were socializing forces. Which was the most important role? Which sex or individual was the most important within the group? These seemed impossible questions to answer, and might not even be appropriate ones to ask of the baboon society I was observing.

The major differences between males and females were equally fascinating. Described in straightforward terms, they seemed simple: types of relationships and sequences of interactions were different. But what it boiled down to was something more ephemeral, both psychologically and emotionally. Males and females, even within the same troop, seemed to live in different worlds. Males were part of a dynamic system, challenging, testing and trying to resolve what appeared to be unresolvable relationships with one another. Among the males, nothing stayed the same for very long; stability was a goal that all seemed intent upon but none ever captured. The only predictable characteristic in male relationships appeared to be their unpredictability.

By comparison, females were dull. Their lives were well ordered and predictable; dominance rank, family and friendships governed the outcome of female interactions. I wondered why the females bothered putting up a fight, since they knew the outcome even before a dispute started. This was the most interesting aspect of female behavior; although everything seemed predetermined, females still tried to maneuver.

At their core, males and females were opposites: males were dynamic, females were stable; the outcome of interactions between males was unpredictable, between females extremely predictable. Males were willing to take risks; females were conservative. Both systems, dynamic and conservative, coexisted within the troop, but they also intersected during sexual encounters and in friendships, where a kind of compromise was reached. Friendships between males and females were much stronger and longer lasting than male–male relationships, but they were much shorter and more ephemeral than female friendships.

---

It was a remarkable set of findings. No male dominance. The reduced effectiveness of aggression to obtain what one wanted. Complementarity of roles within the troop for both males and females, yet a wide divergence in psychological and emotional propensities. Females certainly had an elevated place in Pumphouse when compared with the early descriptions of baboons that had had such an impact on anthropological thinking. The new picture was anything but simple. Much as dogmatic feminists might relish evidence of a primeval matriarchy where female triumphs over male, the baboons pointed in another direction: complementary equality. Just how complementary and how equal I had yet to discover, but the basic outline was obvious.

I suspected that no one would be happy with my findings—neither the feminist anthropologists now spotlighted by the media nor their adversaries, the original interpreters of baboon society. There was no guarantee that I would even be believed. If my initial findings were accurate, the implications would extend beyond baboons to interpretations of the evolution of our own human behavior. If we believe the old position—which took baboons to be the best model of a savannah primate/hominid—or the newer argument that there were principles to be discovered about how any primate could possibly live on the savannah, the same conclusion was inevitable after observing the Pumphouse baboons: aggression, male dominance and male monopoly of the political arena are not necessary aspects of the life-style of the earliest humans. If, on the other hand, we sincerely believe that life, for humans, is based on aggressive competition, male hierarchical relations and male domination, then we may have to come up with new theories and new answers for why this should be so. We can no longer simply say that it is "the natural order of social life."

I had come to Kenya to study both male and female baboons, giving special attention to the "neglected" females. Now I found myself captivated by the males. Merely describing their behavior was not enough, even if the evidence blew apart many old arguments. I wanted to understand their unique mentality. I wanted to be able to predict what a male would do with the same accuracy that I could now apply to a female. I wanted to understand why a male behaves the way he does. Until I resolved the paradox of the enigmatic males, I felt that I would not truly understand what was important in baboon society, and what was trivial—to the baboons, that is, not to their human observers.

# 6. Starting on the Males

I knew what the males were *not* doing, and I could describe what they *were* doing, but I did not have any explanation for *why*. What did dominance mean? What were the uses and advantages of aggression? Why did males spend so much time and effort making friends with females and infants? Why did a male move so cautiously when he transferred into a new troop?

I had other questions, too: Why wasn't I seeing more violence? When the baboons did fight, why weren't more males injured? I could tell from watching Peggy's family that males exerted what could only be called self-control; I could see how restraint was developed during play, and the same type of control was noticeable in the occasional more serious fights among adults.

Paul, Peggy's oldest son, was especially hair-raising to watch in play. He liked to join groups of smaller baboons in wild free-for-alls in which three or four monkeys would gang up on one. But just when the situation started to get serious for the victim, they'd switch to another target. Once in a while the overexcited youngsters would get carried away and the playful melees would turn aggressive, but since everyone wanted to continue playing, the aggression was usually controlled and short-lived. Important skills were learned in these mock fights, and it was in this playful arena that dominance ranks were disputed and established among the adolescent males.

Paul had to display extraordinary restraint when he played with his brother Patrick. Coming full speed from one corner would be Paul, thirty-three pounds of firm, solid muscle; in the other would crouch Patrick, who at that time weighed barely six and a half pounds, and whose biggest "muscle" was his penis! Paul roared up to the infant, then screeched to a halt, contorting himself in order to get down to Patrick's level. Paul held his strength in check for nearly fifteen minutes as he wrestled with his little brother. Such self-control on the part of a hormonally active male was typical.

This gave me a new idea. Would studying adolescent males help solve the mystery of the adults? These immature animals were in transition, moving from the conservative, predictable, well-ordered female system to the dynamic and to me unpredictable male system. How, exactly, did they make this switch? What were the overriding principles that governed their behavior?

---

Before I began to study the males in earnest, I went through a number of life changes myself. I had returned to California in January 1974, worked on my thesis and in September obtained a teaching position at the University of California in San Diego. After visiting my family and friends there, I returned to Kenya in the summer of 1975.

Coming back to Kekopey was like a rendezvous with a lover I'd never expected to see again. It wasn't that the dreams I'd had in which the baboons hung banners from the trees reading "Welcome back, Shirley! We missed you!" came true. It was better than that; the animals continued to ignore me for the most part, and I was able to travel inconspicuously through the troop just as I had earlier, identifying old friends, marking the growth of youngsters and discovering new babies that had been born in my absence.

My visit was short and at the end of the summer I returned to California and my teaching responsibilities. But I left the baboons under the watchful eyes of Hugh and Perry Gilmore, graduate students from the University of Pennsylvania. Hugh was doing his doctoral research on baboon communication.

———————

Now, in July of 1976, I was back with the baboons and ready to face the challenge of the males head-on. I not only had to decide whom to watch but how to watch. In my previous stints, I had switched from animal to animal, spending thirty minutes watching one, then moving on to the next on my list. Such a schedule meant that I couldn't devote more time to any one individual than to another; nor could I give in to the temptation of just watching whatever exciting event was happening at any given moment. This careful control of my observations gave me confidence that when it came time to compare the behavior of individuals, the results would not be contaminated by unconscious personal biases. But there was too much going on to be completely narrow-minded. I appeased my curiosity and added more information to my study by supplementing my main observations with a variety of other techniques that could be carried out simultaneously.

Jumping from one baboon to another, as I had done before, allowed me to watch each one about once every three or four days, obtaining a representative cross-section of every animal's behavior in the morning, at midday, in the late afternoon and early evening. This was a good technique for getting a fix on the basic behavioral patterns and the relationships of a large number of animals, but it wasn't a good way in which to understand any one baboon in depth. If I was going to solve the enigma of the males, I would need a much more profound understanding.

My previous methods allowed me to describe an interaction accurately, but often, with males more than with females, I couldn't really understand its full meaning. I would see the start of something interesting, but by the time the situation resolved itself I was off watching someone else. At other times I would begin to follow an animal in the middle or at the end of what was obviously a much longer interaction. My data contained a great many unrelated beginnings, middles and ends. The time problem loomed large; I knew that baboons had long *memory / history* memories and sensed that there was even a history that went beyond the immediate sequence of behaviors in the more complicated interactions. How could I get at this?

The answer came slowly. I wanted to begin my study of adolescents with a pilot investigation. The results of this mini-project would help me with my final design. I chose Sherlock as my test adolescent male. Sherlock was Anne's son, a member of a middle-ranking family. His age was somewhere between seven and eight. By starting with Sherlock, I could avoid extremes of behavior that might be associated with very high or very low rank. I would later expand my observations in both directions of dominance.

Sherlock was a typical adolescent male. His long limbs and muzzle hinted at what his final size would be, but at this stage he was extremely gangly, lacking about ten pounds of adult muscle, and his adult mantle was just starting to grow. His slightly pointed, upturned nose was a distinctive feature of his smooth, narrow face, and this, together with his extremely light coloration, made him easy to spot, even at a distance.

I enjoyed my first day with Sherlock. I found him at the cliffs in the early morning, and soon fell in with his daily rhythm: bursts of social activity, some feeding and a little rest, interrupted now and then by a brother, sister or playmate wanting to be groomed. The day ended as it had begun, in the company of his mother and siblings. They climbed the cliffs together and settled down cozily on a small ledge. As I watched them, I felt a special satisfaction; I hadn't learned anything new about baboons, but the experience of "a day in the life of Sherlock" had been different from anything I'd done before.

Two weeks later I was still watching Sherlock and imagining the unimaginable. Suppose I did an entire study on just one animal? I was learning so much from Sherlock about how interactions built up, climaxed and then resolved themselves—a process that sometimes took days—that I wondered how I could possibly study *all* the adolescents without losing track of the special insight that came from focusing on one animal day after day.

But my reputation was insufficiently established. My findings were already considered controversial, and I didn't have the courage to stand up to the criticism I knew such a study would induce. I would know more about one baboon than anyone had ever known, but how would I be able to put it into any kind of context? Was Sherlock unique or representative? Did his family's rank influence his options? Did his age? His size? If I restricted my study to Sherlock alone, I would never be able to answer these simple questions.

After spending two weeks with Sherlock, I was convinced that if I was ever going to truly understand males, I would have to take a different approach. I reached a compromise: I would study *pairs* of brothers, one still a juvenile, completely entrenched in the female system, and one an adolescent on his way to becoming an adult. I would spend an entire day with each family—observing the relationship between the brothers and between each male and the entire troop, first from one sibling's perspective, then from the other's: it would almost be like spending the day with only one animal. I would limit the number of pairs I studied so that I could watch them more frequently than if I had taken *all* the adolescent males as my subjects.

I chose my animals carefully. I needed a representative selection, including males from high-, middle- and low-ranking families. I didn't pick them at random—a common research technique—because I wanted to emphasize contrasts. Peggy's sons, Paul and Patrick, were obvious choices. Not only were they from the highest-ranking family in Pumphouse, but, through Peggy, I already knew this family extremely well. I also included Sherlock: he was middle-ranking, and I hated to waste all the information I had on him. Beth's sons made up the lowest-ranking pair of brothers, and there were five more sets of males in between.

From August 1976 to September 1977, I saw each of the adolescent males take on their own fascinating uniqueness. Yet there were patterns, too, and each pattern taught me a lesson about what it means to be a male baboon.

————

Patrick was by now the spitting image of Paul when he himself had been a juvenile. He was approaching the size of an adult female, but otherwise remained the same animal I had observed before, slightly more mature. He shared his family's dominance, and within the family *hierachy in* ranked according to age—dominant to Thea but subordinate to Peb- *the family -* bles. Patrick was Paul's shadow, going wherever he went, trying to do *reverse age* what Paul did. He was sometimes displaced by his older, larger and thus higher-ranking brother, but often he benefited from both Paul's protective presence and from his leftovers. Nineteen seventy-six was a very dry year, and the baboons relied on oniongrass corms for much of their food. But preparing corm sites took considerable time and effort: the surface grass blades had to be tugged out of the rock-hard ground and the earth dug up before the succulent corms could be reached. Paul would appear on the scene just as a low-ranking female had finished the *benefits of the* last of her preparations and was beginning to enjoy a feast. Displacing *association w/ an* her, he reaped the benefits for himself. But before he'd completely *older male* exhausted one site he'd be off to find another victim, thus giving Patrick the chance he'd been waiting for. Although both Paul and Patrick were dominant to all the females in Pumphouse, Paul's large size meant that there were *never* any challenges, whereas Patrick might have to put up a fuss to get what he wanted; it was easier if he simply tagged behind Paul.

Although Patrick was important, it was <u>Paul's</u> transformation into an adult that interested me most. <u>Family attachments still meant a great deal to him</u>. Mother, sisters, brother and nieces filled his social space and were his usual partners. Paul dropped everything to come to his family's defense when they were in trouble, yet at the same time he was breaking away from them. This showed itself in two ways: <u>he was trying to make friends with several adult</u> females, and also attempting <u>to make the adult males take notice of him</u>.

He was having difficulties doing both. Females fled at his approach— and why shouldn't they, considering that up to now his attentions had meant trouble over a feeding site or a defense of his family? It was little wonder that when he made a beeline for Olive, she didn't wait around long enough to see his lipsmacks or side-to-side head shakes, or to hear his appeasement grunts—all friendly signals. With Paul, as with many adolescent males, friendly intentions vanished when they were ignored for long. Being rebuffed by Olive three times in a row was more than he could take; giving vent to his pent-up frustration, he rampaged through the whole troop, threatening and chasing unsuspecting by-standers, mostly females and juveniles.

The first time this happened, only Olive's family and friends mobilized; they put Paul in his place for the time being. The second and third times, the entire troop mobbed Paul. He was pushed beyond the edge of the group by their aggressive sounds, a series of staccato pant-grunts that rose in intensity and volume—a sound known as a "mobbing vocalization." Chastened, Paul sat in isolation, only venturing back when the troop finally ignored him.

*rivalry for female friends*

Making friends was not easy, but slowly, just as Ray had done with Naomi, Paul convinced Olive to accept him. Trouble came from a different corner: Big Sam, Olive's current male friend. When he noticed Paul's progress with Olive, he reined her in, giving Paul little opportunity to develop the friendship. In fact, whenever Paul was near, Big Sam saw to it that Olive was far enough away to make interaction difficult. Paul vacillated. <u>This was a major difference between adolescent and adult males: when obstacles became too difficult, the adoles</u>-

cent had his family to return to, finding reassurance in their acceptance, contact and company.

But each day was punctuated by brief forays away from the family center. The males attracted him. Paul had worked out his relationships with the other immature males—by virtue of his family's rank, he was dominant to all juvenile males—but among the adolescent males, size and age got mixed up with family history. The larger, older and usually stronger males would become dominant over him, even though they came from lower-ranking families.

Newcomers held a special appeal. When a new male arrived in the troop, Paul rushed to check him out. The welcoming committee was the usual odd mixture of infants and adolescents. For the most part, the infants sat a safe distance away, embracing one another reassuringly. Since it was not "acceptable" for an adolescent male like Paul to reveal his inner turmoil in public, he sat at a distance, intently monitoring the new male's every move. When his long-distance come-hither looks were ignored, Paul pulled himself up to his full height and worked up enough courage to actively greet the stranger. Swaggering over (as much as an adolescent can pull off an adult male swagger, lacking both cape and muscle) and then breaking into a quick dash as he neared the male, he tried a "diddle"—a touch to the genitals of the seated new-comer, on this occasion McQueen. Paul was brave; he didn't swerve away at the last moment, but he couldn't quite manage a real "diddle," where the penis is actually pulled or the genitals fondled. He simply brushed the area with his hand and ran off.

Paul watched McQueen closely for five minutes more before trying another greeting. The mixture of signals he sent betrayed his state of conflict: a "fearful" greeting, combining both friendly lipsmacks and agitated grimaces, and "gecks"—single sharp, coughlike sounds—when the male finally reached out to touch him. This was quite enough. Paul returned to his family.

In the following months, Paul greeted each new immigrant as often as he could, but always with hesitancy; if they responded, he was fearful. In contrast, his relationship with the rest of the males in the troop was calm; he shared a relaxed tolerance with those who had been Peggy's friends during his childhood, and they often rested or fed close together without any sign of tension. With the other residents of the troop, however, Paul tried out a new role.

A year earlier, in 1975, when Paul was still a juvenile, he occasionally threatened females while they were with male friends or consorts. Already dominant to the females, he must have been trying to communicate with their partners. Paul was flatly ignored. He continued until he finally got a rise out of someone. The male, tired of being pestered, put Paul in his place. When dominant adolescent males ignored him too, he rushed at them with greater confidence, slapping out as he ran by. It was as if he were saying "Notice me," and sometimes they did.

But now Paul's infrequent and mild harassing was turning into a serious adolescent occupation. He chose times when a male was at his most vulnerable—when he was in consort, and when he was being challenged and harassed by adult males.

Estrous females in consort were often followed by interested males, and Paul joined these. He left the major challenges to the adult males, but was among the first to harass any attempt at copulation. As a consort picked up momentum and copulation attempts increased, Paul might even harass *any* approach the male would make to the female, presumably realizing that such an approach inevitably meant a mount and then a copulation.

Paul's harassment began mildly: a close approach to the consort pair followed by a stare; as tension grew, however, his actions became more serious: pant-grunts were accompanied by aggressive lunges at the couple. He was not a serious threat to the copulating male, but given the tension generated by the close presence of other males and their obvious interest in the female, Paul could tip the balance between a successful and an unsuccessful copulation. He certainly managed either to interrupt or to completely disrupt a great many copulations by adding to the general turmoil that often built up around a mating pair.

Paul timed his interactions with care, seldom challenging an adult male directly, but closely following all the male interactions. Then into the melee he would go, at a moment when there was little to risk, always on the side of the winning males. On such occasions, if you didn't watch closely, it would have been easy to mistake him for an adult.

Paul gave the appearance of assuming a more adult role in the troop in other ways as well. He was often the first one down the cliffs in the morning, marching off determinedly in one direction, trying to be the leader who decided where the troop was to go. But when the others refused to follow him, his resolve failed. He would sit off to one side by himself, looking steadily in the direction *he* wanted to go, while the rest of the troop stalled or even moved the opposite way. Try as he would, Paul could only convince his family and some of Peggy's friends to follow him. When he realized he was in the minority, he would rush to the main front line of movement as if assuming the position of majority leader.

Paul was now beginning to show physical signs of sexual maturity. His testicles had descended, and the clear liquid he had formerly produced when he masturbated—an event related to very little else that was taking place—now had the milky color of adult semen, probably indicating the presence of active sperm. His sexual interest in females increased, and he moved from merely harassing males to actively trying to consort with receptive females. In doing so, Paul suddenly found *sexual maturity* himself catapulted into the world of adults. He had no luck to begin⇒ *adult roles* with: he was <u>unskilled at being a consort male</u>, and the interesting females weren't interested in him at the right times. When relegated to the old role of consort follower, he now did manage a few serious copulations. One had to admire Paul: sometimes his guile made up for his other deficiencies in experience.

The first time I noticed this was when I was watching Roz, an adolescent who had finally reached the right age and hormonal balance to be of interest to adult males. She had a consort partner, three adult male followers—and Paul. Paul never presented a serious challenge, but when the consort male finally turned to face the threats of his adult male followers, Paul found Roz behind a large rock and copulated with her. She seemed willing enough, but was reclaimed by her adult partner within minutes. After that, Paul and Roz had what could only be called a little something on the side. Fortunately for both of them, the troop was spending this particular morning in a rocky area, and hiding was easy, at least for a few minutes at a time.

Paul's sexual behavior as he matured both physically and socially was a good indication of how adult he really was. Big Sam had been unable to prevent the budding friendship between Paul and Olive, a friendship that provided the adolescent with his first real consort. He "put in a reservation" for her—a pattern I later recognized as quite usual among the Pumphouse males. It was an easy step from following his friend Olive when she wasn't sexually receptive to forming a consort with her in the early stages of her swelling. The two types of pursuit were easy to tell apart, and the sexual intentions of the latter were obvious: mounts and copulations. Yet Paul was consorting with Olive much earlier in her cycle than adult male consorts normally did, and copulating with her when her swelling was still very small. He seemed to be trying to make a firm statement about ownership before the important days around ovulation actually arrived. This tactic was only partially successful; he did manage to monopolize Olive much closer to her critical days than before, but he still couldn't hold his own against adult males during the peak of her estrus.

Paul tried putting in a reservation with other females besides Olive. If he started when they were still early in their cycle, his actions went unchallenged by the adults. His behavior remained contradictory, however, because he still showed a certain lack of finesse with those females who weren't his special friends. He would inadvertently displace the female from feeding site after feeding site. The fact that he was bigger, dominant and not a friend meant that his partner avoided him whenever he tried to get close. Sometimes he lipsmacked, grunted and gave come-hither looks to the female, just as in the early stages of courting Olive, and he was extremely attentive. At other times, though, he would allow Olive to groom him while he was in consort with another female, his interest in being the consort male flagging as the grooming relaxed him. Under the circumstances it was hardly surprising that the consort female often looked elsewhere for a more intense involvement. Then, as if suddenly recalling where he was and what he was supposed to be doing, Paul would resume the role as consort male just in time to prevent another male from taking over. Occasionally the appearance of a possible male rival roused him from his reverie so that he could reclaim his female.

When Paul wasn't embroiled with adult males and sexually receptive females, or attempting to decide where the troop would go, he showed great interest in other troops. Interactions between troops varied a great deal, from peaceful mingling, with the two groups appearing as one to an untrained observer, to active aggression between certain members of each troop. Aggression occasionally escalated: one troop would mob the other more or less the way they mobbed a particularly aggressive male from their own group.

"Intertroop encounters" developed with at least three other Kekopey troops on a regular basis: Eburru Cliffs, a large neighboring group whose home range overlapped considerably with Pumphouse's; School Troop, a group about the same size as Pumphouse living to the northwest; and Cripple Troop, so named because of the number of injured and maimed animals in it. Cripple lived to the northeast, where uninsulated 22,000-volt power lines claimed their share of unwary baboons.

*emigration*

Pumphouse encountered other groups—Crater Troop, August Troop, May Troop—but these were seldom long or involved meetings.

Among Paul's priorities, paying attention to other troops, no matter how distant from Pumphouse, ranked high. In fact, his behavior was often my first clue that another troop might be around. Paul stopped feeding, grooming or harassing and concentrated on the other troop. If it moved out of sight over a rise, Paul would adjust his position to find a spot from which to continue his surveillance.

Life for Pumphouse became considerably more exciting when an- *a new troop* other troop was nearby. During peaceful, close-range encounters, it was the adolescent males who first approached the outsiders, greeting them with curiosity; occasionally the younger males of the two troops played together. A brazen adolescent female with a sexual swelling that was still really of little interest to Pumphouse males would seek out potential partners, usually finding that, just as in her home group, only juveniles and young adolescent males showed any interest. An occasional mount or copulation aroused little concern, but when an adult female in estrus ventured anywhere near another troop, all hell broke loose and she would be herded well back into her own group by her consort partner or male friend—or by Paul, who was extremely active in promoting and protecting the integrity of his troop.

Paul, the adolescent, was the perfect example of a male baboon in transition. He had all the right tendencies, but had not quite grown into them. Though his family continued to play an important role in his life, making friends *outside* the family received its fair share of attention and was to some degree successful: Olive could now be counted on as a grooming, resting and traveling partner, and cooperated with Paul's attempts at consorting. Paul's sexual interests were gradually developing, but he still had hurdles to overcome: female attitudes toward him, the success of adult males over him and his own ineptness when maneuvering in the adult realm. Paul's increasing size and experience were, however, finally having some impact on the adult males: at least they occasionally noticed him, though he was far from being a part of their adult constellation. Just where Paul would end up was difficult to guess; his interest in other troops suggested that, given the normal pattern of adolescent males, he would emigrate. But why and how he would leave, only time would tell.

# 7. The Saga of Sherlock

I continued to watch Paul and Patrick closely, at the same time keeping careful records on Sherlock. By March 1977, he had taught me a great deal, both about males and about leaving home.

Sherlock was my favorite adolescent; perhaps the two weeks we'd shared earlier, in August 1976, had forged a special bond between us, because he continued to be my most interesting and challenging subject. Within two months Sherlock left the Pumphouse Gang, and I decided I had to follow him. There were many days when I was his only companion as he roamed Kekopey in search of other troops, each time eventually returning to Pumphouse. The day he made his final decision to leave his natal group, I, too, had to decide whether it was time for me to transfer as well.

Paul and Sherlock resembled each other in many ways. Both had an intense interest in other troops, assumed a vigilant posture when another group appeared and were among the first to investigate when it came closer. Both remained partially tied to their families while trying to start a different life with other male and female adults. It was Sherlock, though, who had made greater strides along the road that compels an adolescent to leave home. At first, he simply moved to the edge of the troop. No one was actually forcing him out; he just seemed to have become a more peripheral member than before. His attention was always elsewhere, always searching for some other troop, in particular the Eburru Cliffs group.

Soon "peripheral" meant "solitary." I would often find Sherlock halfway between the Pumphouse sleeping cliffs and the open grasslands where Eburru Cliffs liked to feed. This solitary existence was actually an illusion; Sherlock never let himself lose sight of one or the other troop. When he followed Eburru Cliffs, it was always at a great distance, with, as the hours passed, many agitated glances at the cliffline where the Pumphouse Gang had last been seen. Sherlock was <u>extremely nervous</u>. When Pumphouse finally returned to its previous night's sleeping site, an excited Sherlock would throw intense greetings to his age mates, Sean and Ian, from a quarter of a mile away.

Sherlock wavered back and forth, not just between roles but between troops. I followed. Watching him try to <u>make up his mind</u> was fascinating: one day he would be hovering on the edge of Pumphouse in self-imposed exile, the next he would be wandering off again. When he returned, he would move right into the middle of Pumphouse activities, taking up his previous life exactly where he'd left it, sitting and grooming with his family, trying to initiate friendships with females and even making a play for one or two receptive ones. Yet the following day he'd take another side trip, this time with a few younger adolescents tagging along in their own peculiar version of an all-male band, Sherlock leading. But what a leader! When the foray ended, Sherlock the Brave rushed back to rejoin Pumphouse, running right into his mother's arms and relaxing for the first time all day. *ambivalence*

Sherlock seemed unable to come to any definite decision, and those around him were confused as well. Even Anne was hesitant at her son's approach if he'd been absent for a few days. Only with his baby sister, Alexandra, and his youngest brother, Alan, did he find a ready welcome, but it was Alexandra and Alan who got Sherlock into the trouble that ultimately seemed to tip the balance.

There was something delightfully touching about Sherlock's behavior with his family. Alexandra enjoyed sliding and jumping on him, and often nestled in his lap; Sherlock always responded, grooming her even when he wasn't asked to do so. They made a marvelous picture. Sherlock dwarfed the tiny black figure of Alexandra, who was half hidden by his hairy bulk, only an ear, eye and nose peeking out to indicate there was a baby there at all. The gentle, relaxed grooming also attracted Alan and Wound, another sister. Of course, Anne was often to be found somewhere nearby.

This peaceful scene would sometimes be disturbed by disputing males. If they came anywhere near Sherlock, he would clutch Alexandra tightly to him, not really to protect her—she was in no danger—but apparently to protect himself. Having Alexandra on his belly was an insurance policy for him: the males would go elsewhere to vent their frustration. The infant also provided a good passport when Sherlock wanted to approach one of the newcomer males. He would move toward the male, with Alexandra nestled comfortably and in complete trust on his stomach; a male approached in this way would sometimes run off, but at other times simply sit still. There was no doubt about it: Alexandra was useful to Sherlock.

At about the time Alexandra was born, Big Sam revived an old friendship with Anne. Soon Alexandra, too, counted Big Sam as her special male friend, so it wasn't unusual to see him among the cluster of Anne's family. Males don't usually like being too close to one another, but there was mutual tolerance between Big Sam and Sherlock, even though they interacted very little. Then the inevitable finally happened: both Big Sam and Sherlock made a dash for Alexandra at the same time, wanting the special protection that holding the baby could bring.

But who had prior right to Alexandra, her brother or her male friend? Certainly they wouldn't fight over her. The first time this happened, Sherlock claimed Alexandra before Big Sam could reach her but Big Sam had other avenues. His previous tolerance of Sherlock disappeared. Whenever he was with Anne and Alexandra, his simple, direct threats made Sherlock understand he was not welcome.

To make matters worse, Alan seemed increasingly prone to get himself into difficulties, first with one youngster, then with another. Sherlock always came to his defense, but often found himself face to face with the older brother or adult male friend of Alan's antagonist, who was often of larger size and higher status.

Conflict over Alexandra, risky situations because of Alan, not much success in making female friends or in finding consorts—perhaps it *was* time for Sherlock to leave. That he chose Eburru Cliffs as his next home didn't surprise me at all. But his first days with Eburru Cliffs were uncomfortable; he was treated in much the same way I had seen Pumphouse behave to newcomer males. Everyone was tense, and Sherlock's stiff posture and wary attitude betrayed his uneasiness. Bands of infants huddled together and sidled up to him, while he, intent on the adults, either ignored them or—when they uttered sharp, fearful gecks at his closeness—turned benevolent, trying to reassure them that all was well by his grunts and lipsmacks.

Sherlock was not exactly a stranger to Eburru. Eight years of interaction had preceded this event, and although no one in Eburru Cliffs knew his family history or had heard of his recent successes and failures in Pumphouse, his keen interest in other troops made it certain that a number of the Eburru Cliffs' animals had greeted, played and possibly even copulated with him.

*I* was the stranger, yet, as I had learned from past experience, other troops took their cue from Pumphouse when I was around. Sherlock showed no aversion to me, so Eburru Cliffs were no more upset about my distant presence than they were about his. *in the new troop*

All these early approaches to Sherlock were friendly; no one tried to chase him away or to establish dominance over him. Apart from the infants, it was the males of his own age who showed the greatest interest. They would start off to greet him, often losing their nerve at the last minute as he had lost his when approaching new males in Pumphouse. One reached for his genitals, another gave him a come-hither look and yet another almost fell over as he turned to present. Sherlock was reserved; he seemed intensely interested, but also ill at ease.

Everywhere Eburru Cliffs went, Sherlock tagged behind, always at a distance. He fed when they fed, rested when they rested. He was particularly observant when the troop was socializing, stopping whatever he was doing when any baboon approached him. Was he watching, as I had observed Ray watch Pumphouse in those early days, trying to tell by behavior patterns and subtle clues who was related to whom, which animals were friends and which enemies? The troop's interest in him was so keen that some days I wondered if he'd got enough food and rest. He was the last to arrive at the troop's sleeping cliffs at the end of the day, and I wondered how it felt for him to sleep by himself. He was with a group, of course, so he wasn't exactly alone, but in

Pumphouse his small sleeping ledge had been crowded, with Wound and Alan, Anne and Alexandra all huddled close together. Do baboons understand loneliness?

A week passed. As I watched Sherlock, I began to identify and name the members of the Eburru Cliffs troop. It was a large group of over 120 animals, and identification was daunting. Yet I knew I had to make a start; if I didn't, I'd never make sense of what was happening to Sherlock. Did only certain baboons come to greet him, or was it a random assortment? Was his reaction different depending on who greeted him? Was theirs?

By the end of three weeks, everyone was relaxing. The infants and adolescent males greeted Sherlock less frequently. Sherlock inched closer to the center of the troop, and so did I.

Suddenly, females started approaching Sherlock. I hurriedly tried to distinguish them so that I could make sense of the new developments, but before I could give each a name, it became obvious that they were all young and receptive. Each female would approach Sherlock a little uncertainly, turn her swollen, bright pink bottom toward him and then wait, glancing nervously back at him. Sherlock paid polite attention to the swellings, but really seemed more captivated simply by the presence of the female. When no mount or copulation was forthcoming, the young female would run off, sometimes returning a second time, just in case. It was Ray all over again.

Sherlock became a definite attraction for the estrous females in the days that followed; it was not merely the young ones without consort partners who were interested. One after another of the females paraded in front of him, and he refused all invitations, simply exchanging polite greetings. The males of Eburru Cliffs raised some objections to so much attention being paid to a strange male.

Since I knew so few of the troop, I could have been confused by all the activity, except that what I was witnessing was familiar. Male A, as I then called him, walked a female away from Sherlock, time and time again preventing her from getting too close to the new male. Another female was even held in check by a juvenile male, something I had not observed in Pumphouse; clearly, certain males had a vested interest in keeping their female friends, or relatives, away from the newcomer.

Basically the adult males ignored Sherlock, and he avoided them. His only interaction was with Virgil, a male who had transferred to Pumphouse for just a month in 1973 before deciding to return to Eburru Cliffs.

98

This could have been either because Sherlock "knew" Virgil better than any of the other males, or because Virgil was one of the youngest adults. But even this greeting was cut short when Virgil, having caught sight of me, ran off.

Over the next few months, Sherlock continued to insinuate himself into the troop. His first major social triumph came when he formed a friendship with Louise, who had been the first female to approach him. Unlike Naomi with Ray, she seemed keen to become Sherlock's friend. Even so, Sherlock eased the path with greetings, grooming and invitations to approach. Fortunately the interaction with Virgil was only temporarily delayed by my interference. Greeting succeeded greeting as Sherlock followed Virgil through the troop. Was he using this male as a shield, a means to enter the new group, or was he simply trying to work out some relationship with him? Whatever the purpose, it was Sherlock, not Virgil, who finally ducked out.

Interesting as these first contacts with males and females of Sherlock's adopted troop were, I had already observed similar interactions in Pumphouse. It was important to know that Pumphouse wasn't unique, but still, by far the most fascinating part of Sherlock's existence as a transferred male came during interactions with other troops, especially when that other troop was Pumphouse.

I spent my preassigned day once a week with Sherlock in Eburru Cliffs, but I always took the opportunity to watch him whenever I was with Pumphouse and Eburru Cliffs was close. This was a period of frequent contact between the two troops, and for several weeks I had daily opportunities to keep track of Sherlock's progress.

At first, to my surprise, when the two troops came within sight of each other, Sherlock hid. No matter where he was in Eburru Cliffs, when he glimpsed Pumphouse he ran, rushed or sometimes even slunk as far into the rear of his new troop as he could manage without actually leaving it. It wasn't that he was uninterested. Far from it; even at a distance, his eyes remained firmly fixed on everything Pumphouse was doing.

Even more interesting was his reaction to his own family. Big Sam herded Anne away from any confrontation between the two troops; this was the common pattern for male and female friends, and usually the female went where she was supposed to go without much objection. It was hard for me to tell what happened on one particular occasion; it could have been that the nearness of several other males frightened Anne, or that Big Sam was more intent for the same reason. Whatever it was, Anne screamed, which brought Alan and Wound to her side. They couldn't do much without Sherlock's help, but they rallied around their mother anyway. For an instant, Sherlock lunged forward as if propelled by both emotion and habit; the next second he stopped dead, holding himself in check. Did he recognize that changing troops meant shifting allegiance permanently? Would he suffer a setback in Eburru Cliffs if he went to the defense of his mother in Pumphouse?

Why was Sherlock avoiding his old troop? Was he still caught between conflicting desires? Would he return to Pumphouse? Was he trying to make a statement to the members of Eburru Cliffs who might doubt his reliability as one of their own? He also avoided the other troops that Eburru Cliffs encountered, but none so much as Pumphouse. When two Pumphouse brothers, Handel and juvenile Mike, got up enough courage to approach Sherlock in Eburru Cliffs, he tried to pretend they didn't exist, and looked away. They didn't take the hint, and the closer they approached the more nervous Sherlock became, finally actually running away from these two younger and subordinate males.

Months passed. By June 1977, Sherlock had assumed a new role. Heretofore he had been the most timid male in Eburru Cliffs; now he was the most brazen and aggressive in intertroop encounters. The moment another troop was spotted, he went into action, becoming the primary "herder" of his troop's females. While other males attended to their particular female friends, making certain that they didn't stray too near the other troop, Sherlock was all over the place, chasing unattached females or even attached ones whose male friend hadn't become aware of the presence of the enemy. This herding was quite a difficult task, because many of the females, particularly the adolescent and young adults, and the sexually receptive ones, were considerably attracted to other troops.

Nevertheless, Sherlock's frenetic activity wasn't winning him any friends in the Eburru Cliffs troop. Many females were already frightened because he was a foreigner. The "new-male phenomenon"—my term for the general attraction of females when a male first arrives—had worn off. His habit of herding females certainly made the situation worse. Most of the time he simply rounded one up by chasing her in a wide circle, back into a more central position within Eburru Cliffs. It was a little like a cowboy trying to lasso a cow without a rope. Sherlock could never be sure the female would stay in place, and frequently she didn't; while he dashed after the next straying female, the one he'd chased earlier often sneaked back toward the other troop. Sometimes simple chasing wasn't enough; disobedient females were threatened, even nipped, to bring home the point. Then the situation would get even more complicated. While there seemed to be no objection by the troop to Sherlock's herding activities, once he frightened a female, her family and allies came to her defense and he was treated like an enemy rather than as the troop's staunchest supporter.

Poor Sherlock! It was a never-ending job, and there didn't seem to be any rewards. Why the sudden shift from recluse to guardian? Why ~~?~~ persist? It was no way to make friends with Eburru Cliffs, his new troop. What was he trying to prove?

Two more events in Sherlock's life impressed me, and helped to lay the cornerstone for what would eventually become my theory about male baboons.

Sherlock wasn't the only male to transfer to Eburru Cliffs from Pumphouse that year. Ian, the elder son of another middle-ranking female, arrived seven months later, at the end of April 1977. Sherlock had been dominant to Ian in Pumphouse, not by virtue of their family ranks but because he was older and bigger. During the week in which Ian vacillated about whether or not to join Eburru Cliffs, Handel, a younger adolescent from the next-lower-ranking family to Ian's, followed him out of Pumphouse and into Eburru Cliffs.

Somewhat naïvely, I had expected Sherlock would welcome these old playmates from Pumphouse, but nothing of the sort happened: Sherlock was even more nervous around Ian and Handel than he had been around the Eburru Cliffs males when he had first transferred. Despite being intensely interested in where they went and what they did, he avoided them, while on their part Ian was the initiator and Handel the junior partner. They were not brothers; at most they might have been cousins, since I suspected their mothers might be sisters, but there was no way I could tell for certain.

Over the next few weeks, a pattern developed. Ian and Handel behaved like outcast onlookers, just as Sherlock had before them. But this time it was Sherlock who monitored them from a discreet distance. When they did interact, the outcome was puzzling. <u>Sherlock seemed</u> <u>subordinate to Ian, even afraid of him, a clear reversal of the relation-</u> <u>ship they'd had in Pumphouse.</u> He was more interested in Ian than in Handel, but when Handel came close enough to make a statement,

*[margin annotation: dominance reversal in the new troop]*

Sherlock avoided him as well. Yet there had been no fights, injuries or sudden bursts of growth and strength that could have caused these adolescents to reverse their dominance ranks.

Another piece of the puzzle fell into place when Handel left and then returned. Now not only Sherlock avoided him; Ian did, too, a reversal of *both* the Pumphouse and the Eburru Cliffs relationships. Could dominance be tied in some way to the length of time a male stayed with the troop? If so, these males were telling me that it was always the most recent male newcomer who was dominant, not the male who had been with the troop the longest. But why?

Sherlock was finally settling in. He devoted much time to his new female friends and began approaching a few males as well. His relationship with Virgil was unlike anything I'd ever witnessed. The two of them synchronized their feeding activities: from his position behind Virgil, Sherlock monitored many of the other males, sometimes shedding his bodyguard to greet Cyclops, Boz or Aeneas, then rushing back again.

The second important event was really more of a gradual change in behavior. Louise was Sherlock's first female friend in the new troop, but not his last. By this time I had watched the formation of so many friendships that Sherlock's actions did not surprise me. His winning over a few females made him more appealing to others. By now he was even consorting successfully with receptive adult females. Not only did he "put in a reservation" with Louise, he even managed to keep her all to himself. He had similar successes with Justine and Andromeda. He was certainly doing a lot better with Eburru Cliffs than he had in Pumphouse. Was this just the addition of time, experience, maturity— something that would have happened whether he had stayed in Pumphouse or left? Or did being with a new troop do something special for him? One thing was certain: Sherlock didn't want Handel to share in his progress. He demonstrated this by being strikingly possessive of his females when Handel was around. His behavior reminded me of the way adult males prevented newcomers from having access to their female friends. Perhaps Sherlock had finally made the transition to adulthood.

Sherlock, Paul and Patrick, Handel and Mike, Ian and Hoppy, Hank and Benjy, Tim and Nigel had taught me that the path to adulthood resembled an obstacle course. Family ties had to be severed before other relationships could change. As long as a male's first priorities were in being a son and a brother, females remained reluctant to accept his friendly overtures, apparently uncertain whether his approach was really the aggressive defense of his family or the actual initiation of a friendship.

Certainly establishing friendships with females did seem important. The few consorts adolescent males had had in Pumphouse were with their newly established female friends. In the home troop, the adult males proved to be the most difficult hurdle. They were much more interested in an adolescent who had managed to form and maintain a consort, because it changed him into a competitor. The closer in size to an adult, the greater an adolescent's impact. Yet his position was still problematic. No matter how tough he seemed, any adolescent was afraid of all the males who had newly transferred into the troop. With the others, he might be able to work out his relationship: mutual tolerance and a little contention with some; active opposition, with instability in actual rank, with others.

On the other hand, an adolescent who left his home troop seemed to have more options, as if in leaving he was wiping the slate clean. Transferring from one troop into another meant losing a great deal—family, playmates, perhaps even some friends and certainly a predictable and known social world. However, there would no longer be any uncertainty about a conflict of interests, no family to defend against potential friends and no history of aggression toward females he might now want to court. In fact, the newcomer might actually have an advantage with females. Being the new male in the troop would definitely prove attractive to some of them, and initial encounters could develop into real attachments once the novelty had worn off, as Sherlock had learned. Friendships led to consorts, or so it seemed, both in Pumphouse and Eburru Cliffs, and consorts were part of what it meant to be an adult.

Of course, even if the females regarded the adolescent newcomer as *almost* an adult, he was still left to cope with the males. Here Sherlock fared only marginally better in Eburru Cliffs than he had in Pumphouse. Most of the adult males ignored him, and he, on his part, was wary of all of them except Virgil. Ian, Handel, Tim, Mike, Hank and Paul fared no better to begin with, either. Like someone on holiday whose past need not intrude on the present, and who is restricted only by the range of his courage and imagination from creating an entirely new persona, an adolescent who transfers to another troop can become whoever he wishes, providing he possesses the necessary confidence and social skills.

Sherlock had been with Eburru Cliffs for nearly a year when one day I found him alone, watching Pumphouse from under a nearby tree. The troop noticed him too, and responded as if he were a total stranger. It was a far cry from the time Paul had returned after being away for a few days, when they had greeted him with welcoming grunts and he had rushed straight into Olive's arms, the two of them embracing as I had never seen grown baboons do before. This time, juveniles on the lookout jumped to high perches and some adolescents made ready to approach, while adult males herded females away from potential trouble. But this was *Sherlock,* not a stranger. This nearly full-grown male had been born and raised in Pumphouse; his mother, sisters and brother were still there, as well as old friends and playmates. Had they forgotten him? How could baboons, with their remarkable memories, for whom interactions might span days and even weeks, forget someone who had been an integral part of their lives for at least eight years? By what methods and for what reasons do baboons forget?

Whatever the answers, it was apparent to me—and I assumed to Sherlock as well—that it would not be easy to go home again.

# 8. Bo and David

I was certainly learning about the adolescent males and about what it meant to be part of the adult male system, which seemed infinitely more complex than I had imagined. Each breakthrough in understanding was accompanied by a new and puzzling riddle of behavior.

It was now 1978, six years since my initial visit to Kenya, and every so often, just before falling asleep, I would remember my first glimpse of that old truck's motto: No hurry in Africa. There was no irony in this. I wasn't dallying, and the baboons were being marvelously cooperative, but I sometimes felt it would take me a decade to finish my work with the males.

And what about my other life, my life in California with my family and friends? I spent my trips back there trying to explain my ideas and philosophy to my parents, who felt they'd lost a daughter and received very little in return; to my friends, who kept worrying about my future and introducing me to eligible men; to my colleagues—and, especially, to myself.

It's important, I'd tell them. It has to do with all of us, with our pasts and our futures. What I'm discovering is important, but it's not easy and it's not quick. Forgive me, I sometimes wanted to say. Understand me.

But why, they asked, couldn't I simply stay in California and study people instead of baboons? I could certainly learn as much, and I could begin to lead a "real" life.

True, I could study people; I'd have a great advantage: I could ask my subjects about what happened at certain times, what they were feeling and thinking, what seemed important or trivial to them. But there would be many disadvantages—not the least being that human subjects can lie. I remember one classic study that spanned nearly a decade. It was five years after the study began that the subjects informed the anthropologist studying them that they had lied to him about everything; now, since they felt they could trust him, they would tell him the truth!

When one studies animals, there are no problems with lying, but there is the problem of how to understand a creature that does not talk. We often forget that we are animals ourselves—that we watch the outside world with specialized senses, with a brain that is geared to integrate this information in a specific way and with a set of emotions strongly invested in one view of how the world works or should work. Many early interpretations of animal behavior were unconsciously anthropomorphic, projections of human behavior onto animals. The problem was greatest in studies of monkeys and apes, since our biological closeness to another creature influences our ability and desire to view it in human terms. It is more difficult to guess what two insects are doing than to intuit the behavior of chimpanzees; and it is even more difficult *not* to assume what is happening when we watch the higher primates, because we *are* so alike and can understand so much more of their communications and emotions.

It was in large part for these reasons that the earliest studies done in this century often went astray. It is easy for us to spot the flaws in E. Kempf's study, published in 1917,[*] in which he concluded that homosexuality was a natural stage in the development of adult human

[*]"The Social and Sexual Behavior of Infra-human Primates, with some Comparable Facts in Human Behavior," *Psychoanalytic Review*, 4:127–154.

sexuality. He watched rhesus macaques that had been caged together specifically for this study. His observations were correct: he saw the adult males mounting one another frequently, ignoring the females that were caged along with them. Ergo, male homosexual monkeys, he concluded. There were two problems with this interpretation. Primates use many social signals that are derived from sexual behavior yet do not necessarily denote sexual interest. Mounting and grasping another male by the hips can be a greeting or a statement of rank: it is rarely actually sexual. Furthermore, the lack of sexual interest that this particular group of monkeys displayed toward their female cagemates could easily be explained: none of the females had reached sexual maturity.

With more time spent observing animals, a better theoretical grounding on which to base questions and more sophisticated ways to record observations, animal watchers gained confidence that they were no longer simply casting their subjects in the human image. European ethologists of the 1930s and later had provided sound rules about how to observe animal behavior, but there were still major obstacles to understanding what the observations meant.

I had inherited these ethological rules, and had followed them to the best of my ability: do not interact with your subjects; do not label behaviors with terms that hint at interpretation; first construct an ethogram, whose units should be small, well described and useful to anyone else who watches these animals; make no value judgments during descriptions and observations; eliminate every possible source of bias from your data.

"Clean" data, educated hunches and verification had brought me a fair distance, but not far enough. Male baboons still baffled me; clearly I was missing something. It was only when I got a closer look at the history of two young males, Bo and David, that I made any further progress. Bo and David were identical in age; one had been a juvenile transfer into Pumphouse, the other a native who had never left. After I had studied them, following them along the road all males in the troop traversed, I found some of my answers. It took me nearly a decade, which, in the end, seemed a triumph.

When Bo came to Pumphouse from Eburru Cliffs in 1973, the idea that his stay would be permanent was so improbable that I dismissed it; he was only about four years old at the time and clearly was only going to be around for a few days. After he'd grown up, perhaps he would choose Pumphouse for his home, but in the meantime he'd return to his natal troop. So I told myself. I was wrong. Bo had followed two males from Eburru Cliffs into Pumphouse. I called them Strider and Brutus, the first from Tolkien's *Ring* trilogy and the second because the name seemed appropriate to a male who acted like a brute. All three—Strider, Brutus and Bo—stayed with Pumphouse.*

As a juvenile, Bo received little attention from the Pumphouse adults. As long as he behaved well he was left alone: he must not frighten infants or interrupt interactions, and he must keep a proper distance at all times. This was familiar to him: the same rules apply in any baboon troop.

Bo was interested in playmates and grooming partners—friends, in other words—but not in grown-up friendships that involved sexual and social complexities. The males of his own age accepted him with only slight hesitation. The females appeared a little more concerned, but soon Bo found himself in the midst of juvenile activities. He was fearful and hesitant of the larger adolescent males, even though he accepted their play invitations; he actively avoided the adult males.

David was a Pumphouse native, just Bo's age; the two were identical in size, but there the resemblance stopped. David's coat was golden and Bo's was gray; David had a short curved tail, while Bo's jutted out from his body. David's brow was even; Bo's had a slight Oriental slant at the corners of eyes and eyebrows.

I knew nothing about Bo's background: his family's rank in his troop, whether his mother was still alive, how many brothers and sisters he had, what their ages and dispositions were. On the other hand, I knew David well. He was Debbie's† oldest; little Dierdre, alias Dieter, was his sister. A new daughter, Dawn, was born the year Bo arrived, and later Desdemona joined the family. David was involved with both his own and their families. There was nothing distinctive about him; he seemed an average baboon, spending time with family and playmates,

---

*A few of the juveniles to join the troop over the next decade accompanied what appeared to be an older brother, an adolescent or young adult male who had selected Pumphouse as his first home away from his natal troop. However, it is highly unlikely that either Brutus or Strider was related to Bo. They were middle-aged males, too old to be his brother, and given the rate at which adult males transferred, it is also unlikely that either of them was his father. Whatever their relationship in Eburru Cliffs may have been, once the three were in Pumphouse, Bo received no special attention from either of them.

†Debbie was part of a middle-ranking clique of females. Both Marcia and Frieda ranked below her and both were younger. All three had very short tails and a peculiar body build; they could have been related in some way. They were certainly friends.

lending his support to them when necessary and beginning to test his position with higher-ranking females, a prelude to pushing his way through and out of the female hierarchy.

At first there was no special affinity between Bo and David; they played together occasionally, but that was all. When I started my study of adolescent baboons in 1976, both were nearly full-grown physically, although they still showed some signs of immaturity. Since Bo had joined as a juvenile, I thought it would be useful to see how he compared with adolescent transfers and young males born within the troop. David, whom I'd omitted earlier because he'd had no younger brother, would be a good male to pair with Bo, who was about the same age.

I began what was to be a five-year study—1976 to 1981—of the two of them, with each year yielding crucial new data about what it means to be a male baboon. Initially they took different paths, but in the end these paths converged.

Bo had been a well-mannered juvenile, but as an adolescent he turned into an aggressor. He either sat on the edge of the troop like a new male or was stirring up trouble in the thick of it. He bullied females and youngsters, terrifying them into flight. Sometimes a male friend of Bo's victim came to the rescue; then the aggression would spill over. In male-to-male confrontations, he was also on the receiving end of considerable trouble; there were times when the whole troop defended his frightened adversary, rushing at Bo in one great phalanx, screaming and pant-grunting, stopping only when he had been evicted.

As he grew, Bo began to actively incite others. Going straight for an adult male, he shadowed him first, then harassed him until the male could take it no longer. One by one, he conquered each of the adult males in Pumphouse. Soon the entire troop was avoiding him. McQueen, seeing Bo angling toward him, went straight to Kate and began to groom her. By the time Bo arrived, all that greeted him was an unexpressive back. Bo saw many such backs. Getting nowhere with McQueen, he headed for Rad, but Rad loped over to Mary just in time to settle into a grooming position and again Bo was left out. There was always Strider, but he, too, was vigilant, and Peggy made a perfect partner for him. So it went, until, having impelled all the fully adult males to groom a friend or grab a baby, Bo turned his attention to the other adolescents. But they held little challenge—Sherlock started screaming at him from as far away as twenty feet, even though Bo had not yet looked at him.

Aggression took other forms. Although Bo wasn't the newest male in the troop, he became the general herder in intertroop encounters. Like a well-trained collie, he made certain that none of the Pumphouse females went over to the other side. When he herded, females listened; he was greatly feared. He was also a contradiction. He was the most aggressive baboon in Pumphouse, but when a nearby baby screamed—not at him, but at someone else—he darted away as fast as he could. When other males fought near him, even if he had been the one to precipitate the fight, he fled. Here was an aggressor afraid of aggression. He was also an ardent follower of consorts, and here, too, was another contradiction. After days of following a certain consort female and her various partners, he finally could be seen with her himself, early in the morning, coming down from the cliffs. But his consort, instead of lasting hours or days as with the other males, took only a few minutes. This pattern repeated itself over and over again: Bo in consort early in the morning, but only briefly.

These consorts ended strangely. Several interested males would press in close behind the pair. Although Bo was individually dominant to each of them (when he approached one, the other male would get out of the way, screaming in fear), he was obviously nervous. It could have been the number of males that upset Bo; something obviously did. Whatever it was, he either walked off abruptly, leaving the followers scrambling for the female—whoever got there first was usually the next consort male—or else attacked a nearby infant or juvenile, completely without provocation. In the commotion that followed, he would run off. Bo's behavior in the first case was a giveaway; the second seemed a diversion that allowed him to leave without losing face. The un-provoked aggression against the infant was not aimed as a sophisticated tactic; in fact, the baby was usually socially unrelated to the followers. Had they been related, there would have been an even greater problem; although every male was frightened of Bo, fear turned to aggression when a male's friends were in danger.

Bo's relationship with David was unique. They were magnets whose polarities reversed suddenly, frequently and unpredictably. For most baboons, ambivalence goes one way: an adolescent wanting to greet a new male might be both welcoming and fearful; a low-ranking female attracted to a new infant could be frightened of its high-ranking mother. Bo and David were mutually ambivalent, and the result was an extremely unstable relationship. The on-again, off-again interaction

between Ray and Big Sam, from which I had first learned about the
dynamic instability within male relationships, seemed almost static in
comparison. Bo and David were always watching each other, sometimes sending greetings across the troop: eyes narrowed, chin pulled in,
head shaking, one would invite the other to approach. Instead, he often
ran farther away.

It was difficult to tell how Bo and David ranked with each other. One
day David seemed to avoid Bo, the next it was the other way around;
sometimes they'd even switch roles in the course of the same day. David
always kept well away from Bo when he was socializing—probably a
wise move, since most of Bo's interactions were aggressive. Once Bo
departed, however, David lost no time in moving in.

David was Bo's only "friendly" social partner, if you could call their
ambivalent relationship friendly. Except for the times when he was
with David, Bo seemed to be trying to fight his way into the social
center of the troop, using up a great deal of energy but not getting very
far. He *was* dominant to all the adult males, even if their ranks with
each other were uncertain or unstable. But if dominance was to have
any meaning, shouldn't it lead to the possession of something valuable
or necessary? I was unclear about how Bo connected with his early-
morning consorts; I was certain he possessed the female for such a short
time that he was unlikely to have sired any offspring—presumably the
real purpose behind the effort. Why did he lose the female? He was
dominant to all the followers, so why should he be nervous? What role
did the female play in his attitude and failure? Bo was still socially
peripheral. By winning status he lost friends; therefore his rank gave
him only negative influence.

David was completely different. He was the only male who never left
his natal troop in all the years I watched the baboons. Given what I was
learning about why adolescent males left, he was a fascinating study.
Life was tough for him. It was extremely hard for David to make the
transition to adulthood in a troop where he had been defending his
family throughout the course of his life, and had been cast in the role
of aggressor toward females he now wished to befriend. How could he
get other males to notice him, until recently a small, subordinate baboon who was part of the "female system"?

David's early approaches to females outside his own family met with
the same reluctance, resistance and fear that each adolescent male in
Pumphouse encountered. As with Paul, frustration led to aggression:

he even attacked the female he had just tried to appease, a tactic that made the females fear him, though not as much as they did Bo, who was *always* aggressive, never friendly.

It is difficult to tell whether baboons are neurotic in the same way humans are. David, however, came closest to what one might imagine *the neurotic baboon* a neurotic baboon would be like. He took failure badly; when females rebuffed him, he would lurk behind bushes and rocks.

Gradually, almost imperceptibly, David distanced himself from his family. When a crisis developed, he was slow to arrive on the scene and when he did get there he defended his mother and sisters only half-heartedly, often leaving before the crisis was settled. Then he began to *estrangement from family* ignore his family's calls for help completely. Instead of sitting with them during the day, grooming and being groomed, he kept away. Whether this was deliberate or accidental I never knew. By now he was sleeping alone. I came to realize that if I had just begun to study the troop, it would be hard for me to tell that David was a member of a Pumphouse family. There were no clues left that would help identify his mother or sisters.

While Bo was flexing his muscles, David was trying to socialize within the troop. He lacked finesse, but persistence eventually got him the female friends he aimed for. It wasn't easy. Females either rejected him outright or merely tolerated him, obviously preferring someone else. David remained undaunted. Intent on helping Mary, a young female he had been following for weeks, he attacked her aggressor, only to have them both turn on him. Even this didn't discourage him. At other times, Mary wasn't above using David. When he was around, she bullied higher-ranking females, confident that David's protective aura would shield her from retribution.

Fortunately, David was more successful with other females. First there was Diana, a young adult with a small family. His subsequent friendship with Diana's baby, Desiree, was nearly as strong as Diana's maternal bond. After Diana came Beth, the lowest-ranking female in the troop, then Tina and her infant Tito, the newest member of this sizable middle-ranking family. Finally, while juggling as many of his existing friendships as he could manage—the association with Diana was clearly suffering—he took off after Vicki, one of the oldest females in the group and fairly high-ranking.

Female resistance was one thing, but some of the females already had male friends who were not happy about David's moves. Here David—

like Bo—was a contradiction. Although he was much more sociable than Bo, he could be aggressive, and sometimes responded with blind rage when he was thwarted. Frustration meant aggression, usually toward anyone nearby. Like Bo, he could press home an aggressive advantage, at least with some of the adult males, but even more than Bo he avoided aggression in anyone else. Screaming infants, fighting juveniles or females, and grappling adult males were definitely to be avoided. Although some of the time David himself caused the conflicts, he seemed wary of involvements that could lead to trouble.

However, male objections didn't deter David: they only slowed him down. While Bo used up all his energy on conflicts, David appeared to direct his to greeting, appeasing and approaching females.

At the same time that David was trying to win friends, he was also playing around with consorts, though at first these sexual encounters were as brief as Bo's had been. He was visibly more nervous than Bo; either he would "give away" the female, or else use Bo's diversionary aggressive tactics to extricate himself.

While Bo was stuck in a rut, David did make progress. First he tried putting in a reservation, hoping this would be sufficient to hold a female during the key days of her cycle. However, he succeeded at this only with newly won friends like Tina and Vicki. Even when there were many male followers, these friendly consort females showed little interest in them, and would stay close to David rather than straying, as his other consort females had. Once he had a friendly female in tow, David didn't take flight when followers offered their inevitable challenges; he simply stuck with his female and watched as the aggression was deflected away from him to the followers themselves.

David was learning a few tricks. During one particularly hotly contested consort with Tina, he managed to walk her up to the top of the insulated high-tension electricity lines that bisected Kekopey. The followers found it impossible to reach him on his high perch, and discord broke out in their ranks.

Had David's special relationships, his friendships, paid off? Consorts with female friends had many advantages, and certainly he appeared more confident. He was able to stand up to challenges and it seemed easier for him to hold on to his female. A friendly consort partner stayed nearby when squabbles broke out, leaving him free to devote his full attention to the followers. Even in calmer times, there seemed to be payoffs. With other consorts, it was the male's job to groom the female (if she deigns to groom *him*, the time spent is short and the grooming superficial), but David's friendly consort partners returned the grooming they received. Perhaps just as important, now that he no longer had to worry about his females straying, David actually found the time to feed as well as to copulate.

These friendships also benefited the females. David was quick to defend his new associates—not just the female, but her family as well. Because of this, and as if to complete his transformation, his own family sometimes became the opponents in his indefatigable friendship campaign. The battles were mostly trivial, but the first time he actually attacked his own sisters in support of one of his friends must have come as quite a shock to them. Eventually the message sank in: David was no longer a member of his biological family. They avoided him, treating him like a stranger, a male of unknown origin new to the troop. Meanwhile, his recently acquired friends and their families began to treat him as if he were a blood relative.

Bo and David, two males so alike in some ways and so different in others, shared the inexperience of all adolescents. Both were easily frustrated when they couldn't achieve their goals, and for both frustration often led to aggression. Rampaging through the troop usually brought the same response: irate troop members joined forces to mob the offending male.

But to my eyes, their differences were overwhelming. Bo chose an aggressive path and seemed determined to fight his way into the group's social life. Yet for all his aggressive dominance, he appeared to have achieved little. He was a high-ranking male, but he saw those of lower rank get what he desired: special resources, friends, estrous females or the meat from baboon kills. He was a social outcast, feared by all except perhaps David, and even he was ambivalent.

David, on the other hand, had chosen a social avenue. He was dedicating his life to making female friends. In the early stages, the troop's females seemed to prefer nonnatal males, but David's perseverance overcame their reluctance. Although he wasn't as individually dominant as Bo, he was where Bo wanted to be—right in the center.

Over the next few years, between 1979 and 1981, David continued to avoid his family and to accumulate females; he had more female and infant friends than any other male in Pumphouse. His relationship with the rest of the males stabilized. While he was dominant to most, he seldom pressed the point and often joined in coalitions with them. When he was faced with aggression, he had a number of options. He could try to enlist other males in his defense, or he could use one of his infant friends or even a female as a buffer. (See figures 1–5.) He began his transition to adulthood with a strategy of affiliation, not aggression, and stayed on that path with great success.

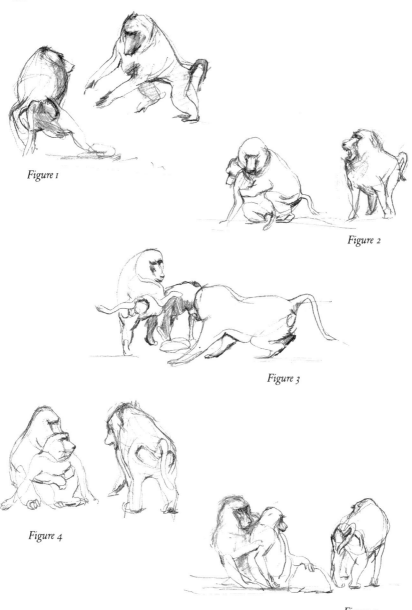

Figure 1

Figure 2

Figure 3

Figure 4

Figure 5

Bo, too, pursued his initial approach—aggression—until a serious foot injury transformed him. He was no longer able to harass those around him, and as a result switched from aggression to affiliation. To my surprise, he showed even more social finesse than David. His added age and maturity may have allowed him to obtain with skill what David had achieved by persistence. Although he had a lot of catching up to do, both with friendships and consorts, he nonetheless may have had greater potential than David. Females who were not yet tied to any male seemed to prefer Bo over David. Was it because he was not born in the troop?

Bo and David had taken two different approaches to the problem of becoming adult. The two roads were not equally successful. Bo's change of tactics was convincing evidence that aggression was not a good passport into the social world of these baboons, neither as an entry permit nor as a visitor's pass. Watching Ray in the early days of my first study in 1972–74 had suggested the same conclusion. Despite his strange proclivity to remain with his natal troop, David's behavior paralleled Ray's, but it was really Bo who convinced me that not only was dominance (when it could be determined) not what it should have been, but that aggression, too, was different from what I and others had been led to imagine.

# 9. Some Solutions

At last, in July 1979, I turned to the data I had been collecting over the years on fully adult males. I had gone as far as I could with the adolescents and juveniles; now I would examine *all* the males covered by my eight-year study period.

Many new males had entered Pumphouse; some of the original ones had left for other troops and some had simply disappeared or died. I'd had the opportunity to watch seventeen males closely, certain ones for only the few months they'd lived with Pumphouse, but others had stayed much longer, some the entire period. Ray, Rad, Sumner, Carl, Arthur, Big Sam, Strider, Brutus, Virgil, Dr. Bob, Gargantua, McQueen, Reynard, Duncan, Angus, Higgins, Chumley and Antonio were my subjects. Each baboon was a unique character, yet there were factors even more important than personality. The patterns of male behavior seemed to be related to the length of time spent with Pumphouse. Size, age and weight paled in comparison with the residency factor, but rank was also implicated in a peculiar way. I shuffled and reshuffled the cards containing the life history of each male. Everything I needed was there, but how did it all fit together? I lined the males up, first one way, then another.

One combination seemed to work better than the rest. I tested it again, and again it worked: The Pumphouse males fell into three categories of residency within the troop: *newcomers, short-term residents and long-term residents*. I assigned a cutoff point for each category; moving the boundaries a little to one side or the other didn't seem to make much difference. Life is continuous; it is only those who seek to analyze it who need discrete categories. All males except David began as newcomers. Some stayed, and after a year and a half became short-term residents. If a male stuck it out for at least three years, he could be considered a long-term resident.

I was now ready to tackle the first issue: *dominance*. Dominance rankings, although not stable or linear, could be predicted from residency class. Newcomers were the most dominant males in the troop, then the short-term residents, with the long-term residents last. This conclusion seemed counterintuitive, since heretofore it had been assumed that a male's dominance was based on his size, strength, biology or fighting skills, not his tenure.

I took a deep breath, and turned to *aggression*. Aggression was also a function of residency. Newcomers were very aggressive males, short-term residents less so and long-term residents were almost devoid of aggression. Again my results placed me at odds with earlier ideas. Aggression was meant to be functional. Males presumably used it when they wanted to achieve a desired goal or needed to protect themselves. It should not matter how long a male had been with a troop: when the occasion arose it was expected that he use aggression—or so I had been taught. Another deep breath.

Finally, who received the rewards was supposed to be determined by dominance ranks. But these baboons were determined to get me into trouble. It wasn't the newcomers, the aggressive and dominant males, who had the greatest success with females or in getting favorite foods, as everyone would predict. The most successful males were the long-term residents of Pumphouse: the lowest-ranking and least aggressive individuals. The least successful were the aggressive newcomers, while the short-term residents fell somewhere between the two extremes.

I tried to think the whole pattern through again. Newcomers were high-ranking, aggressive males who received very little; long-term residents were low-ranking, relatively unaggressive males who got the lion's share. I would never be believed unless I found reasons to justify such an unusual set of relationships. As I reviewed more events, I gained confidence. I considered the adolescents—Sherlock, Handel and Ian. Although Sherlock was dominant to both of them, and Ian was dominant to Handel in Pumphouse, once the three were in Eburru Cliffs their relationships changed. Sherlock, hardly a long-term resident but still the senior of the three in their adopted troop, became subordinant to Ian. When Handel arrived, he became the dominant member of the three. Each ranking accurately reflected the length of their residency in Eburru Cliffs. Sherlock, the longest, was lowest-ranking, Ian was next in both tenure and rank and Handel, the newcomers' newcomer, was dominant.

Something stronger than a hint came from the behavior of two males who moved between Pumphouse and Cripple Troop. Sterling, the Cripple Troop adult arriving in Pumphouse, was, as a newcomer, both aggressive and dominant to the resident males. After testing the waters for a few weeks, he returned to Cripple. A Pumphouse male, Mike, soon followed him there—at which point their roles reversed, as did both their dominance and aggression. Sterling, only a few days earlier a feared adversary who had been avoided and screamed at, became the subordinant resident male, screaming and avoiding the newcomer. Had the two males undergone a sudden personality change? Were they the product of an instantaneous genetic mutation? Had Sterling from Cripple Troop lost his physical vigor and the Pumphouse male come into his prime in a matter of days? This was highly unlikely, considering the equally sudden reversal that occurred when the two males returned to Pumphouse. Residency status reversed once more, as did everything else.

Residency, dominance and aggression were all inextricably linked; there could be no other conclusion.

But why didn't dominance and aggression get a male what he wanted? Why was there an inverse relationship? I took a closer look at consorts. Males sometimes fought over a female, but once a male had her he should be able to dominate her by size alone. Yet watching David, Paul, Sherlock, Bo and others had shown me that females played a definite role in consort success. David had been at his most successful when he was in consort with his female friends, and this appeared to be true of other males. A friendly female made life easier; a male had fewer obstacles to overcome when feeding, traveling, grooming and copulating. Just as David had shown more confidence in facing follower males when he was with a female friend, so other males were better able to resist the pressure of followers when *their* partner was a friend, someone who would stay close and not run away.

There were times, however, when females *did* exercise some choice. I remembered how Zelda kept running from Antonio. There was no mistaking her intentions as she headed for a thicket and zigzagged through it, hiding from him. This was a last resort; she had tried losing him all day, but none of the rocks or bushy outcroppings were big enough to hide her. The thicket was the answer, and it worked. Not only did Zelda shake Antonio, but Gargantua, a favorite friend, was on the far side of the thicket. Had he been waiting for Zelda, or was his presence just an accident? Whichever, Gargantua was Zelda's next consort, and the difference in tone and behavior would have told any observer that they were friends.

At other times, females simply wore out any male they didn't particularly like. Distinct from the giveaway that Bo and David used in similar situations, the male left the female after a harried and unproductive attempt to keep up with her. Was he saying that it just wasn't worth it after all?

Friendly females in consort also behaved differently in many subtle ways. I couldn't say just why they did this yet, but the fact that they did so was important. Having female friends seemed critical to a male's ultimate reproductive success. Perhaps this was why Ray was so intent on making friends, and why David, Sherlock, Paul and eventually even Bo pursued this goal so intensely.

The possibility also existed that males contested a consort less if the male and female partners were friends. Hans Kummer has demonstrated* that among hamadryas baboons, who live in harem groups, once a male gains a female, other males are inhibited from challenging his right to her. There were hints of this among Pumphouse baboons, and the situation deserved closer examination.

Even when males were aggressive, the results were not overly dramatic. Only 25 percent of the cases studied showed aggression as being the reason one male won an estrous female from another. More often,

*H. Kummer. "Dominance versus Possession: An Experiment in Hamadryas Baboons," in E. Menzel, ed., *Precultural Primate Behavior*. Basel: Karger, 1973. H. Kummer, W. Goetz, and W. Angst. "Triadic Differentiation: An Inhibitory Process Protecting Pair Bonds in Baboons." *Behaviour*, 49:62–87, 1974.

devious tactics won the day. For example, Sumner might sit on the sidelines at a distance, watching a consort pair and its male followers eagerly but discreetly—until, before anyone knew what had happened, he was the new consort partner. What *had* happened? I watched such sequences several times before I could determine the exact order in which events took place. Sumner would wait, biding his time until the tension became too much for the consort male and he turned on the followers. While the combatants were disputing, Sumner would dash in and walk the female away. Possession can be nine points of the law among baboons.

Sometimes the deviousness became more active, as when Sumner incited aggression from the ranks. As the fight heated up, he would duck out and claim the female while the others were occupied.

But this subterfuge was modest compared to other maneuvers. Vicki, in consort with Strider, had been avoiding her particularly aggressive followers all day. I watched as the group went around in circles. No one made any progress—not the consort male doggedly following Vicki, not the followers, who were unable to oust him, and not the troop, who had spent much of the day following the consort group. Now, as they were all resting in a thicket, Vicki's daughter Vanessa made a beeline for her mother, hoping to be groomed. Suddenly Rad intercepted Vanessa, frightening her more than actually attacking her, and both mother and daughter fled. Vicki hadn't even seen what had happened, but seemed shocked into action by her daughter's screams, and Rad, right behind them, became the new consort male. Confusion reigned among the followers. They looked around for the culprit, exchanging halfhearted eyelid flashes and pant-grunts, then set off in pursuit of the new consort pair.

Sumner and Rad were not the only devious baboons; all the long-term residents behaved the same way. But in the wider scheme of things, short-term residents were more devious than newcomers, who almost never used such tactics. Success in getting a consort female, keeping her and perhaps even copulating with her seldom resulted from aggression. Rather, Pumphouse baboons used *nonaggressive social strategies* to achieve their aims. Aggression is, of course, one kind of social strategy, but the idea that the nonaggressive alternatives could be as important and as effective as aggression in situations where males competed for critical resources could be a revolutionary idea.

Why were these nonaggressive social strategies used so *unequally* by the males? Eventually it became obvious. To succeed in a social manipulation required understanding and insight into the social networks and complexity of the troop; it needed experience and skill, and it demanded that a set of social relationships already be in place. No wonder newcomers seldom used such tactics and that long-term residents had an advantage. Limited in their options, losing important contests to long-term residents, newcomers seemed to abandon aggression as fast as they could. With time and effort, new avenues opened for them: friendships with females and infants, knowledge about the other troop members. Once these avenues had been explored, a male could try finesse instead of force.

With the exception of David, every male began as a friendless, ignorant and relatively unsophisticated newcomer, and the degree of ignorance and gaucherie depended to some extent on his age. Mature males had years of experience, albeit in other troops, and achieved their goals in Pumphouse more quickly than younger ones in the same situation. Initially, friendless newcomers did attract sexual interest in young resident females, but the new-male phenomenon wore off quickly, leaving the individual feared and avoided. A newcomer probably discerned relationships and the finer details of group life by watching others, but watching was not enough; integration into the troop required action. And act the newcomer did, on two fronts: by befriending females and harassing resident males. Ray was a classic newcomer. The type of interactions he initiated with Naomi, Peggy, Big Sam and Sumner were constantly repeated as other newcomers tried to penetrate the troop.

If a newcomer didn't give up (and some did, leaving Pumphouse to return to their previous home or to try their luck with another troop), in time he became a resident. At this point, a short-term resident. It seemed to take about a year and a half for a male to become fully integrated into Pumphouse. The newcomer would metamorphose, making friends and reverting from aggressive posturing. He also lost the dominance that being feared had brought. These changes did bring about their own rewards—greater consort success, for example. Female friends would cooperate, and the male would have accumulated a certain degree of social smartness. Aquired social skills would, in turn, help short-term residents obtain other valuable items.

After being resident in the same troop for more than three years, a baboon became a long-term resident. In the world of baboons, I was discovering, this meant that he had *made* it. Long-term residents, with

*integration*

many friends and a considerable number of successful consorts, were the most socially skillful of all the males. Low dominance and infrequent aggression ran parallel with the fact that when this type of male *did* act aggressively, such aggression was well timed, effective and a tactical feat rather than a disastrous rampage.

*Residency*

Gradually I gathered even more evidence that residency was the key to understanding males, and that it was the existence of social strategies that helped explain why dominance and aggression were of little use to a powerful male baboon.

One of the disadvantages of being a long-term resident was that the newcomers were always aggravating. Just what was all this harassment and aggression about? I remembered when the aggressive Ray won his encounter with Big Sam when the latter was in consort. Although it had looked as if the males were competing over the female, Ray had walked off, showing no interest. If his aggression wasn't competitive, just what was its purpose? Such aggression seemed to be a way of testing resident males—of "working out" a relationship. Just as juvenile males pushed their way through the female hierarchy, or as adolescent males tried to change their status in the eyes of the adults, adult males often used aggression solely to get another male's attention.

Aggression *said* something. We had always assumed that when a large male employed it he meant "I want what you have," "Get away from me" or some equally belligerent statement. But lacking human language, how else could a male even begin a negotiation with someone who wished to ignore him? How else could a male discover what he needed to know about other males? In responding to aggression—and it was difficult not to, in one way or another—Pumphouse residents gave away a great many secrets. To a newcomer, a resident male was a stranger with no past and an uncertain future, so predicting what he might do was difficult. For a resident, being an enigma had its advantages, and refusing to interact with a newcomer left all his options open.

What could the residents do when a newcomer could no longer be held at bay? I had expected them to meet aggression head-on, either calling the bluff or giving ground. Sometimes it did happen this way, but frequently—as when Ray pushed Big Sam a little too often and a little too hard—the resident sidestepped the issue by using an infant or female as a buffer.

Using infants as buffers against another male's aggression had been widely observed in other baboon groups studied throughout Africa, but it remained unclear as to exactly why putting a baby on your belly would make an aggressive male subside. Of course there were as many possible answers as there were investigators. Some thought that the infant's black color—other monkeys also have babies that are a different color from the adults—was developed during primate evolution to protect a baby from aggressive attacks; they suggested that baboons reacted strongly and instantaneously to the black color, and aggressiveness would stop immediately. An adult male who grabbed an infant would be benefiting from this evolutionary development, using it to his own advantage. But within the Pumphouse troop, black infants were not immune from attack; if the tactic didn't work for the babies, why should it work at second hand for the males who used them? The situation was complicated further by the fact that Pumphouse males also used brown infants—older babies that had already turned the adult color. Such infants lacked the "special" black color; what possible protection could they offer the males?

Another set of interpretations suggested that males were able to recognize which baboon was the infant's father. I had my doubts, but kept an open mind. In one scenario, males forcibly kidnapped their opponent's offspring. Holding the aggressor's "reproductive success" captive should give him pause. But although I examined all the information on who mated with whom and who could have fathered a specific infant, the interpretation didn't fit. It was the Pumphouse residents who used infants against newcomers, and often these newcomers hadn't even been with the troop when the infants were conceived.

Yet another version of the story turned this theory topsy-turvy: males were not using infants as buffers; they were protecting them from being harmed by infanticidal newcomers. In this case, the infants were supposed to be related to the protecting male, not to the newcomer. As usual, the Pumphouse baboons didn't fit in with this interpretation, either. Most of the time, a Pumphouse male would get an infant from somewhere else and bring it close to an aggressive newcomer who, until then, hadn't expressed any interest in it. This hardly seemed an effective way to protect an infant, or a good explanation of why the newcomer should stop being aggressive.

I puzzled through my own explanation. Resident males were using infants in a special way, primarily against newcomers. This happened in a number of situations, ranging from simple nervous tension to actual aggression. Using an infant as a buffer did not always prove successful. I found an important clue: when an infant buffer cooperated with an adult male, there was a much greater chance that the aggressor would back down. When the infant wriggled, tried to run away and was generally uncooperative, not only did the opponent continue to be aggressive, but—if the infant was sufficiently upset—the troop might even mob the buffering male.

Which infants cooperated? Infant *friends.* Friendship meant trust and cooperation, which in turn could produce success in using an infant to ward off the aggression of another male. No wonder newcomers didn't use infants; they had no youngster they could depend upon to cooperate with them because as yet they had no friends. Trying to pick up an uncooperative youngster could generate serious trouble. But very young babies who were *not* friends would sometimes be more cooperative than an older infant who wasn't a friend. Perhaps because they were not socially experienced, young babies put a naïve trust in adults.

Using a baby to ward off another male's aggression was also related to the response of the troop. When my adolescent male subjects became too aggressive, the troop often unleashed its full fury, mobbing and chasing them to the periphery of the group and beyond. The same was true of aggressive newcomers; Pumphouse defended infants and females—but not adult males—against their abuses. So if a male wanted the power of the troop behind him, he had to get hold of an infant as an insurance policy against both defeat and injury.

Yet holding an infant served a more complicated end than that of mere insurance for the male. In subtle ways the presence of the baby could make a difference in how the male felt and acted—and this, in turn, could change the odds. A baboon learns from infancy about the reassuring comfort that contact with another monkey can provide, whether mother, brother, sister, playmate or friend. In fact, the actual relationship didn't really matter; what did matter was the contact. The calming effect of contact was obvious: one need only look at mother and babies, grooming pairs, friends or families sitting quietly together after a frightening time. Because of their size and status, adult males had a limited range of contact partners, and infant friends were the most important.

There was no way to know what actually happened within a male once he picked up an infant, and my hunches were based on what I observed. Agitated males became quiet; insecure males seemed more confident. Wouldn't these changes have an effect on the opponent? Wouldn't *I* respond differently to someone who seemed relaxed and confident rather than overcome by fear? Wouldn't *I* view my opponent differently if I were calm rather than agitated?

*calming effect*

I was venturing into uncharted territory, but it did seem logical and reasonable that just holding an infant could change both how the male viewed his opponent and how his opponent viewed him. And of course there was also the insurance-policy aspect.

A few comical cases showed yet another way being with an infant could solve the problem of aggression. Antagonists sometimes reached a stalemate from which neither seemed able to extricate himself. Suddenly there would be a baby in the picture, and all eyes would turn toward it as timid resident and fierce newcomer both grunted and lipsmacked for all they were worth. Soon one or both males would walk off as if they had conveniently forgotten their original dispute.

As I recapitulated my years of male baboon studies, I grew more and more sure that social strategies, not aggression, were the ingredients of male success. One type of social strategy was useful in competition, as between males for sexually receptive females; the other was useful in defense, as when a male used an infant or female as a buffer.

Social strategies had to be engineered and learned, and that was why the long-term males were the most successful. Newcomers had few options because they lacked both social ties and experience. Their aggression—one of the few options they *did* have—made them feared, and thus dominant, but neither dominance nor aggression gave them access to much of what they wanted. Short-term residents were on the way up; they had made friends and gathered social information, but it was the long-term residents who showed how much time and experience was needed in this male world. They were the lowest-ranking, least aggressive and most successful. They had wisdom, friendships and an understanding of the subtle tactics necessary. They had the greatest potential to maneuver and manipulate, and they did.

The more I understood males, the clearer it seemed that they had a hard life. Where did their size, strength and physical power get them? They were constantly reenacting their own expulsion from their Garden of Eden; they had to construct whole new lives for themselves after leaving behind all that was friendly, secure and familiar. Once in their new troop, they had to recapture all they had lost:

the social closeness, allies, friends, experience and knowledge that ultimately constituted baboon wisdom. Despite great effort and enduring patience, they ended up with less than a female who had never left home could command simply with a look, gesture or grunt. I felt sorry for males.

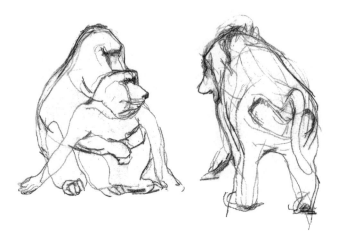

# 10. Smart Baboons

S mart baboons. Everywhere I looked I saw socially smart animals. As social diplomats, they employed finesse—social strategies, sophisticated maneuvers and reciprocity—for their very survival. I had discovered extraordinary intelligence, planning and insight in their interactions with one another, both as individuals and as members of real or constructed families.

Just how smart were they? What were the limits to these newly uncovered dimensions of the "minds" of baboons? Some answers to these questions came as I reviewed a remarkable incident that demonstrated group learning: Pumphouse becoming sophisticated predators, real hunters. Bob Harding, Pumphouse's first observer, had already documented that the baboons killed and ate hares, young antelope and birds. But the troop had been opportunistic then; it neither hunted down the animals in a coherent, premeditated fashion, nor did it scavenge for dead animals; the adult males simply collected the prey they came across, usually young animals hidden in thickets or tall grass.

In 1973, during my first year, Pumphouse became truly predatory, and did so rather suddenly. One of the reasons was that Kekopey was unusual because there were few dangerous predators around, either as potential baboon killers or as competitors for prey. Kekopey was also the home of just the right species—Thomson's gazelles. These "tommies" were an appropriate size and lived in herds that made them easier to find than other secretive smaller antelopes like dik-dik and steinbok.

Until 1973, Pumphouse was no more predatory than their Amboseli National Park counterparts, who lived in constant danger from big cats and other predators. I first noticed the change in the females, who seemed more interested and involved in eating meat when one of the Pumphouse males had a prey. The youngsters joined in, snatching up scraps from the little that was left by their mothers. Then females began to try for their own prey, lunging at hares and antelopes.

But the most impressive change was in the behavior of the males. It happened over a period of months, and it was Rad alone who was primarily responsible. He was a young male with a keen interest in meat; up to this point, Carl had been the most persistent hunter, and had exercised an overbearing influence on Rad from early on, excluding him from many opportunities. Then Carl suffered a severe injury to his arm, and Rad came into his own. Unencumbered, he had a field day, adding his own individual touch to baboon predation.

Rad was like Naomi, hanging around the edge of the troop but not really peripheral to its workings. Unlike Naomi, however, he didn't mind lagging behind or rushing ahead. He appeared less tied to the troop than most males—certainly less than Carl or Sumner, the main predators. Soon he began leaving the troop to observe a nearby herd of Thomson's gazelles, hoping to find a young fawn hidden in the grass. After catching it and enjoying his meal in splendid isolation, he would return to Pumphouse covered with blood, so that there could be no mistaking what he'd been up to.

The other males were aware of what Rad was doing, but at first took no action; they simply watched him leave and return. As he became more successful, however, they began to pay more attention, shifting to better positions so as to observe his activities more carefully. Then they joined in. At first it wasn't to try and help Rad; each had his own selfish intentions.

Then an incident occurred that seemed to change the way the males hunted. Until that point, even when several of them went after a tommy baby, they chased it in any direction, often out onto the open plain. On this occasion, Rad closed in on a group of tommies, scattering the animals, surprising a mother and fawn and almost succeeding in grabbing the baby. Then he chased in earnest, running at full speed, trying to get close enough for a second lunge. Over the hill came the rest of the males. Rad was at the end of his endurance; baboons can run fast, but not for any length of time. Just as Rad gave up, Sumner took over, with Big Sam and Brutus also converging on the prey. The chase turned into a relay race, one male running after the fawn and another taking over when the first one tired. Finally, Big Sam chased the young antelope right into Brutus's arms.

The baboons obviously learned from this experience; frequently thereafter I witnessed hunting in which one or more of the males chased their prey toward another hunter, and their success rate climbed rapidly.

Further elaborations followed. Females were trying to capture their own prey and youngsters were learning to eat meat by observing and imitating their mothers or male friends. Some of the youngsters were even trying their hands at getting a live animal for themselves as long as they didn't have to stray from the troop to do so. When one male juvenile wrestled with a baby tommy close to his own size, he flipped it over, only to have it flip *him* over, continuing this way until Rad claimed the fawn for himself. (Juveniles did manage to capture hares and birds by themselves.)

The primary prey for the adults was young tommies, and the baboons were becoming more discriminating. Earlier, a male might follow any herd, and since tommy groups come in three forms—mixed herds, all-male herds and female and young herds—this wasn't always productive. By now, the male hunters had learned to survey the herds, selecting for their predatory forays only groups that might have babies. Both the length of time spent on hunting and the distance covered increased markedly, until it wasn't unusual for a hunting trip to cover two miles and take two hours from start to capture. To complicate the situation, the tommies were changing their behavior as well, beginning to treat the baboons as the true danger they were instead of ignoring them. Pumphouse had to work harder and harder to get close enough to have a chance at a fawn.

Throughout these changes, the males and females behaved dramatically differently. No matter how keen an interest a female had in eating meat, she wouldn't leave the troop to find prey. Even Peggy, the most enthusiastic and most dominant, stayed within the troop. When males made a capture and brought it closer, females and youngsters would rush to the carcass and eventually have their turn at it. But if the troop moved off while they were still eating, first the youngsters, then the females and even some of the adult males would rush to join the disappearing monkeys, often leaving a half-consumed carcass. It was easy to understand why. Females stood to risk a lot if they were caught alone. They were also usually encumbered by some stage of childbearing, and had little mobility or speed. Males, on the other hand, even though attached to the troop, were leavers. Their impressive size and fighting equipment put them in a different class when they did follow their migratory inclinations.

Sometimes the male-female differences in behavior over predation and meat were comical. Once a female had had an opportunity to eat meat, she valued it above everything else. Sumner in consort with Peggy was a classic case. Peggy stared fixedly at a nearby carcass as Sumner copulated with her. When he was done, she determinedly circled back to the carcass—which was surrounded by males—while again and again Sumner tried to chase her in the opposite direction. On the other hand, a male like Sumner always preferred sex to eating meat. His deliberations were transparent as he glanced back and forth between the kill and his female, hesitating only briefly before continuing with his consort.

The baboons were changing fast. The predatory rate climbed: In 1973, a hundred kills were noted in twelve hundred hours of observation, not much by the standard of traditional human hunters like the Bushmen, but almost ten times the rate of the Gombe Stream chimpanzees. In terms of real meat consumed, the baboons were probably doing even better, since their prey was heftier than chimpanzee prey.

But it was not merely the techniques and rates of predation that were changing. The baboons were beginning to *share* their kill—a previously unheard-of development.

Baboons never share food. A mother won't even share food with her infant. The two might eat from the same patch, but no giving took place. Chimpanzees *did* share food: a discovery that challenged one of the bastions of human uniqueness. Now the Pumphouse baboons had become sharers. No one ever simply handed the meat over, but a male would scoot aside to allow a female friend a turn at the carcass, or a mother would let her infant join in the meal. As usual, Peggy was in the forefront. She exploited her good relations with Sumner and Carl, two of the key predators, and was the first to see they would share. Sumner did so graciously, but Peggy always chose the exact moment that he looked away to grab her big pieces.

Surprisingly, none of the males ever shared with one another, not even with males who might have wittingly or unwittingly helped in a chase and capture.

Surprising also in this increased predatory activity was that the baboons still ignored opportunities to scavenge. This fascinated me, and I conducted a series of small field tests to try to help understand why. I had noticed that not all the animals, not even all the adult males, knew how to get at the brain of a gazelle. There were several techniques: cracking the skull between their molars, piercing it with their canines or even eating through the upper palate to the bottom of the braincase, where it was softest. Limited baboon expertise meant that a carcass was often abandoned with the brain still remaining.

After making certain that no one was looking, I picked up the remains of a kill—pelt, limb bones and head still attached. When I caught up with the troop, I presented the carcass to a number of animals. The results were enlightening. Only those baboons who had actually seen the animal captured or who had been among the original consumers would accept the carcass and treat it as something edible; the rest were afraid of it and ran off to a safe distance. Even more interesting was that once a single baboon had accepted the carcass as food, the others, even those who had originally rejected it, changed their minds and jockeyed to get a turn.

I repeated this with a number of carcasses, finally becoming convinced that baboons took their cues from one another. This could be important. Baboons are not true carnivores, and might not be adapted to handle the diseases that accompany already-dead animals. The baboons seemed to follow a rule: If a baboon had witnessed the killing, or if another one felt that the meat was safe to eat, then it was safe for everyone. If the origin of the carcass was unknown, or if there was reluctance on the part of other baboons to touch it, it should be avoided.

From the troop's enthusiasm and success I had assumed that hunting would become a permanent part of its behavior. I had thought I had witnessed an evolutionary step forward—if "forward" means "more like humans"—but the Pumphouse Gang's predatory sophistication turned out not to be permanent. As time went on, males left or disappeared. The new males were more interested in one another and in females than in meat. By the late 1970s, although the females and youngsters kept up their end, continuing to eat as much meat as they could, the sophisticated hunting of the adult males had been erased. None of the remaining baboons, even if they happened to be potential hunters, had experienced at first hand the new hunting techniques I had observed earlier. Although there were still as many tommies available, there were no more intensive searches, long stalks, relay chases. Tommy fawns were now relatively safe, and the occasional antelope, hare or bird that got caught and eaten was just unlucky.

---

Baboons were smart, not just in the way that they had developed predatory techniques, but in other ways as well.

I had only to think of Peggy and Dr. Bob's tommy carcasses for the idea of baboon "smartness" to take on a new dimension. Earlier, Peggy had managed to get Sumner to share his meat with her; later, when the males had changed and one male, Dr. Bob, was reluctant to share, she outmaneuvered him. The first time she lulled him into a grooming stupor before stealing the tommy carcass from his lap. The next time, only a few days later, Dr. Bob was wiser. He let Peggy groom him, but when she made the least movement toward the carcass his hand went firmly down to hold on to his prey. Peggy would get away with stealing once but no more. Thwarted but not defeated, she looked around until she found Dr. Bob's closest female friend: Peggy herself was a friend, but not as high on Dr. Bob's list as this female. Deserting Dr. Bob, Peggy rushed to the female and in a very un-Peggy-like manner attacked her without provocation. Dr. Bob was in a quandary. He *should* go to support his friend, at least to reassure her after such an attack, but it was too far to drag the carcass. He looked back and forth between the female and the meat, obviously trying to make up his mind. Finally, deserting the meat, he started toward his friend—passing Peggy, going in the other direction on her way to the carcass!

There were plenty of similar examples of baboon cleverness. Olive, a young female, was being harassed by Toby, Tina's adolescent son. He was indulging in the usual aspiring male pastime of attacking females in order to improve his rank within the troop, and this time had chosen Olive. He challenged her on the far edge of the troop, where her screams went unheeded; it was windy, and they probably could not be heard. Olive avoided Toby, conceding her rank in the process.

When the pair rejoined the troop, Toby was dominant to Olive. In the natural order of baboon existence, they would remain this way for the rest of their lives together. I followed Toby for the remainder of the day, and was still with him at noon, when he again ended up close to Olive. Their reunion seemed to be accidental. Toby was feeding, not really watching where he was going. Olive glanced up: there was Toby—and there, too, was her older brother, Sean, who was bigger than Toby and dominant to him. She looked intensely at Toby, then at her older brother, then back again at Toby. Suddenly she screamed bloody murder. Simultaneously, the two males looked up. Her brother's actions indicated that only one conclusion was possible: Toby was the culprit and Olive was the victim, and Sean was not going to stand for it.

When the chase ended several minutes later, both males had run the distance of the troop and far beyond, with Toby the loser. The dominance ranks reversed again, although it was inevitable that eventually Toby would again rise in rank over Olive. Had Olive been settling a score, trying to get even, was she using her brother to keep her dominant status a little longer? Whatever her intentions, her screams were unwarranted and the consequences were to her benefit. She seemed to be manipulating her social environment with considerable dexterity—something I myself could manage only infrequently.

Social skills like these had to be learned and perfected, and I had had ample opportunities to watch how they developed. I knew that immigrant males had to learn the ropes, how to maneuver through the social networks of families and friends. Youngsters born in the troop did too. Thelma was one example. She was still an infant and not yet independently dominant when she was displaced by Tina. A few minutes later, seeing Peggy walking nearby, Thelma tried to incite Peggy against Tina, wanting to use her grandmother's rank to her own ends. But Peggy would have none of it. In her estimation, the situation didn't warrant the commitment Thelma wanted. It was in ways like this that young baboons learned both social skills and how and when to push the limits.

Then there was the whole issue of friendship. Tim Ransom had identified friendships among the Gombe Stream baboons between males and infants and between males and females. He had hinted at ways that these might be useful to the various partners, but never implied any tactical awareness or foresight by the baboons. Pumphouse demonstrated a larger and more impressive phenomenon to me. Friendships were almost formal systems of social reciprocity. The underlying understanding seemed to be, "If I do something good for you now, you'll do something good for me later." The balance sheet would be set up in an individual's favor by a combination of good deeds and hard-won trust. This was quite a sophisticated process when one took into account the time that might pass between credits and debits. A new male coming into Pumphouse acted as if he had thought to himself: "To be successful in this troop, I'll need a few female friends, several infant friends (preferably at least one tiny black infant) and some male allies." He would then set out to acquire them. Weeks, even months later, he would call in his dues, relying on the cooperation of a female friend during a consort, counting on the complicity of an infant during agonistic buffering and depending on the help of one particular male in the face of a daunting male opponent. I now knew why infants and females cooperated with male friends. It was enlightened self-interest. In their unwritten contract, these social partners traded what they had to offer for other benefits: a male friend's assistance during critical social situations. This included both active physical protection and indirect consequences of staying close, which could forfend bullying. Just sitting near a male friend improved access to special foods, better resting places and other limited resources.

Male social sophistication had other facets. Agonistic buffering— placing another baboon between oneself and an aggressive opponent— was a major feature of male life. Juvenile males occasionally used their younger siblings as buffers, learning some social ins and outs in the process. Adolescent males, too, used young family members as agonistic buffers until they started to form friendships with females and then with unrelated infants, finally blossoming into the adult male pattern. Adults didn't use infants and females indiscriminately as buffers. Infants were more effective in some situations than in others, and the same was true of females. What mattered was who *you* were, and who your opponent was. The longer your adversary had been with the troop, the better an infant was at turning off his aggression. By contrast, newcomers reacted more strongly when confronted with a female buffer, especially if she was high-ranking. This made sense. Newcomers wanted to have female friends. Being seriously aggressive toward a female, even one that has been brought into the fray by another male, is not the way to begin or to promote a friendship. The longer a male resides in the troop, the more friends he has already made, the more socially established he is and the less he needs to be concerned about what impression he might create. What *is* important to him is safeguarding his friends and offspring. But it is often difficult for a male in a promiscuous species like baboons to recognize his own offspring. If, as a resident, he has reproduced successfully—one of the characteristics of residents—that infant on his opponent's belly might actually be one of his own children. Therefore it was best not to take chances.

It had taken me months poring over the information on agonistic buffering—information that spanned a decade and numerous males—to reach these conclusions. Did the males keep the rules in their heads? If they did, then each time two males interacted, each one integrated and evaluated a lot of information and experience, far more than we had ever given them credit for before.

I had hints of something else just as remarkable. After watching Pumphouse for thirteen years, I was to recognize that some males had stayed long enough to become *longest*-term residents, living with the troop for more than five years; one actually stayed for ten. I was curious to see if there was any pattern to the way in which the males left Pumphouse to join other troops. There was a big exodus after about one year of residence, mostly of the males who had been unable to integrate themselves into the troop and who left to try their luck elsewhere.

Then there was a marked lull in emigration, followed by another exit of males after about five years of tenure. This was a magical figure for anyone seeking biological explanations. At that point, if a male had been reproductively successful, his daughters would just be entering the pool of sexually competent females. Could these adult males be leaving in order to avoid incest and inbreeding? This was the evolutionary reason offered for an adolescent male's first move away from his natal troop. Certainly for as long as adolescents did stay within the troop, they showed little interest in their mother or sisters although they often tried to mate with similar aged, unrelated females. Such observations were provocative, and were consistent with findings from the Cayo Santiago groups of macaques and also with studies of the sexual behavior of people reared in Israeli kibbutzim. Why the selective interest? Only the study of people offered any insights. Children reared together in children's houses had no sexual interest in one another when they grew up, even though most of them were unrelated. Familiarity seems to breed sexual boredom—a good evolutionary device, gone astray in the case of the Israelis, but very adaptive in the case of macaques and baboons.

Were the adult males leaving to avoid inbreeding? If they didn't leave the group at some point, the inbreeding problem could be serious. Even if it was possible for a male to identify his own daughters or to avoid all the females of that age group, as time passed he would find fewer and fewer "safe" females to mate with.

Of course males could leave Pumphouse at the five-year mark without being aware of the evolutionary reasons for doing so. Something as simple as an emotional proclivity, programmed over the millennia, similar to the one operating on adolescent males, could impel departure. A more insightful reasoning might be at work, however: counting up the attractive females and becoming aware of the diminishing choices.

———————

As I pondered the intelligence of baboons—their cleverness, social sophistication and the new intellectual powers such talents might imply—my thoughts kept returning to the stable female core. Although this core reflected a simple linear dominance order of families, females, in fact, showed many contradictions. They were interested in predation but reluctant to attempt anything new or risky. (Later, in 1981, I was to watch another contradiction on Kekopey, at the time the baboons were raiding the crops of local farmers. While the females were initially cautious about doing the actual raiding, they seemed willing to tear their own families asunder in order to follow the raiding males, and in the end they themselves began to take part in the raiding.)

The stable female dominance hierarchy began to show itself as more enigmatic than it had originally appeared. Mothers normally outrank their daughters; the entire family focuses on the matriarch. While a baby is small, the mother's support helps to establish its rank; later, that assistance is no longer needed.

This usual course of events eventually turned topsy-turvy in Peggy's family. She herself became very ill; her infant Pebbles disappeared and was presumed dead; Paul was killed. What actually went through her daughter Thea's mind is impossible to determine, but one day Peggy simply ceased to be dominant to Thea, who pressed home her point by consistently harassing her mother, now too weak to put up a fight. Paul, who would have protected her, was gone, and outsiders seldom meddled in family affairs.

I'm certain that Peggy's poor health and generally depressed state provided the opportunity for Thea to subjugate her. But why did she do it? Was it because Thea really *was* a bitch? Confusion was created within the family. Peggy's granddaughters, always happily subordinate to Peggy and never suffering as a result, didn't understand what was going on. They watched their mother displace and harass their grand-mother; only after weeks did it dawn on them that they, too, were dominant to Peggy. Thereafter, Theadora, Tessa and even little Thelma followed in their mother's footsteps, picking on Peggy unmer-cifully, even violating the first rule of family life and siding with strang-ers against her. Peggy put up little resistance, as gracious in defeat as she had been in victory. Patrick was not so docile. He tried to defend his old position and managed to aid his mother a little, cutting short the harassments of his nieces. The peaceful grooming rituals that had always been this family's hallmark disappeared; aggression within the family skyrocketed. In the end, only Thea was dominant to Patrick. The rank order within Peggy's family, with daughter and granddaugh-ters dominant to mother, and older sister dominant to younger brother, was unique among the troop.

No one has ever seen a female dominance hierarchy develop in the wild. They simply exist, and are remarkably stable. Reversals between females are extremely rare, which indicates that there is something akin to a conservative nature in the female character. However, it is a trait we are unable to measure. A reasonable speculation is that the female hierarchy could have evolved in the following way:

At one time there was one female baboon and one male baboon who were the only members of their group. Soon the female had a baby. A little later she had another and then another, until their group was no longer small. Her children had a rank among themselves that reflected their mother's willingness to support them: the youngest was highest-ranking and so on in reverse rank order, back to the oldest. After many years the daughters reached adulthood, still retaining their birth rank. They began to have children of their own and the cycle repeated itself. The matriarch finally died, but the group continued to grow. All the adult females were neatly ordered in a hierarchy, but as the group increased, many females became only distantly related to one another.

Whatever the real story, what I saw in Pumphouse was that female rankings were usually clear-cut and stable. Even Peggy, once the most powerful female in the troop and even now second only to her daugh-ter, contributed to that stability when she did not try to reclaim her rank after she regained her health. How unlike a male this was! Males seemed to thrive on dynamism, and their dominance ranks were really only small probabilities that one would win over another in a situation where reversals were the rule rather than the exception (a far cry from what scientists normally mean by dominance hierarchy). The rarity of

T. W. Ranson/National Geographic

1

S.C.S.

2

T. W. Ransom

3

1. The Pumphouse Gang on
   Kekopey

2. Peggy, my most important
   guide in Pumphouse

3. Relaxing with the baboons
   at dusk

S.C.S

Neil Leifer

5

S.C.S.

6

T.W. Ransom/National Geographic

7

4. Females and infants
   socializing

5. Older infants ride on their
   mothers' backs

6. Josiah with the baboons

7. Persuasion is often called for

S.C.S

8

8. Male with an uncooperative
   infant

9. Male with infant friend

S.C.S

9

S.C.S

10

T. W. Ransom

11

T. W. Ransom

12

10. Black infant with older play-mate—note the change in color between the two

11. Male eating a baby gazelle

12. The days end at the sleeping cliffs

T. W. Ransom/National Geographic

13. The baboons were fascinated by my Siamese cat

14. The troop in a maize field

15. Educating local people about wildlife

16. Shirley and the baboons

13

S.C.S.

14

S.C.S.

15

B. Campbell

16

17

S.C.S

17. The baboons were captive for a few days prior to translocation

18. Family and friends resting after translocation

19. The translocation didn't change the baboons' attitude toward me

18

S.C.S

19

S.C.S

such reversals among the females underlined the fact that family members seldom took advantage of one another, even when the opportunity to do so existed. Any family member violating this unwritten rule of baboon society did so at her own risk.

The prediction was fulfilled several years later, in 1984, when Thea was overthrown by the eldest of her daughters residing in the troop at that time. She had a bad foot injury that not only hampered her movement but was clearly extremely painful. The exact moment of the rank reversal was not observed, but it couldn't have been very dramatic or it would not have gone unnoticed. It was easy to speculate that suddenly, during an interaction, the daughter simply refused to give way; perhaps a skirmish took place that Thea was physically incapable of winning on her own, and there was no one willing or able to assist her at that moment. The reversed ranks became an established fact thereafter, but a period of some ambiguity followed, straining the normal set of family interactions, just as it had with Thea and Peggy.

I wondered if the pattern would repeat itself in the next generation, and if Theadora's daughter Tootsie would displace her mother as the two earlier generations had done.

----------

The stable female hierarchy sustained other jolts. When a matriarch died, leaving the family headed by an adolescent daughter, this young female might try to push her way into a higher position in the existing hierarchy. These efforts seldom worked, the females ending up with an even lower rank than the one they had before. Was it worth the risk? I wondered. Why did they try?

With all the new evidence I had about baboons' "smartness," I was still puzzled by some of their behavior. The female dominance hierarchy itself posed a particularly knotty problem. Rank implied that there were some advantages to be gained, but in order to be important, what was gained had to immediately or subsequently reflect on individual survival and reproductive success. What exactly were the benefits of high rank for a female? She could push another baboon away from a good feeding or resting place, and she could win other battles or skirmishes, but if their rank was of any real importance, high-ranking females should also produce more babies—and more of these should survive.

I had thirteen years' worth of information on many of the females, a period that covered more than half their reproductive lives. For some, of course, I had it all. The evidence was from good, bad and in-between times—an excellent cross-section. But when I examined the details, I found that high-ranking females neither produced more infants nor had a higher rate of infant survival. The existing pattern resembled something found in other mammals, a pattern that had not yet been documented for a nonhuman primate in the wild: age-specific fertility.

Two factors did influence female output—age and the availability of food. Young and old females were much less fertile than middle-aged females, and produced fewer babies. The oldest females tended toward such a slowed fertility rate that it was likely that, if they lived long enough, they would experience menopause; certainly apes and monkeys in captivity did so. Females of all ages produced fewer babies when there was little food, more babies when they had more food available to them. Compared with the factors of age and nutrition, the effects of rank were insignificant.

One other possibility remained, but it was one I could explore only after all the females in the troop had died. Did high-ranking females simply live longer, producing even one more baby per lifetime? This was all it would take for female dominance to be important. If lower-ranking females were harassed more and were under greater social and nutritional stress, they might die earlier, thus giving high-ranking females an advantage. This is possible, but it remains for me to test the idea over the coming years.

I was confused. All the evidence pointed to baboons being remarkably clever social sophisticates in all aspects of their lives. Females had a clear and pervasive dominance hierarchy, but *why* did they? I knew the reasons why males lacked one. With all the male comings and goings, stable relationships were hard to achieve; more important, dominance rank secured a male relatively little of what he wanted. He needed social strategies, skills and experience that would allow him to finesse his way to his desired goals much more than he needed dominance rank.

The existence of the female dominance hierarchy began to pose an equally formidable challenge to my understanding of baboons. High rank didn't ensure a female's reproductive success. Unlike males, females were reluctant to try to change their rank, except in a few rare cases, such as in Peggy's family, or when the head of a family died. Yet females managed to get around their rank by a variety of means. A good example was Beth, the lowest-ranking female in the troop. She buffered her low position in several ways: she always stuck close to her male friends, and she had many sons, each of which, on reaching adolescence, would be his mother's most ardent and successful defender. Neither of these facts actually changed Beth's rank; they simply cut down on the number of baboons who pushed her around or flexed their status muscles in her vicinity. If high rank provided less stress and with it a longer life, then Beth, a low-ranking female, created a similar existence by having friends and producing sons. Evolutionary biologists had argued that low-ranking females should have daughters and high-ranking females should have sons. Beth's life suggested a novel twist to the argument. But was there really any way that females could influence the sex of their offspring?

Beth also did some adjusting on her own. When Benjy was born, the mother-infant pair became the focus of attention of other females; unfortunately for Beth, all of them were dominant to her. At first she ran away at the slightest sign of their approach; to have let them come close would mean trouble. But slowly she learned that all they wanted was to touch and grunt at the baby; they weren't interested in whatever she herself was eating at the time. Gradually Beth modified her behavior: first she stopped moving so far away from the intruders; then she stopped moving at all; then she simply stopped what she was doing while the dominant baboon was nearby. Finally she became almost blasé about the whole business, munching away while permitting the other female to attend to her infant.

As Benjy grew older and became less attractive to the group's females, an approach often meant trouble rather than a baby-seeking encounter. Beth relearned her old ways and resumed her avoidance techniques. Nowhere in the entire process—and it happened with each of her babies—did she change her dominance rank. She merely learned how to make her social position the least disruptive as possible to her goals. So what was the *real* cost of low rank to Beth? Could she control some of the disadvantages that being subordinate brought? She could obviously use her behavior to maneuver in and out of difficult circumstances, but did her control go beyond that—could she dictate to her body to produce sons? How smart *was* Beth?

I was also struck by female disputes and fights. Since females knew their ranks and the likely outcome of the fights at the start, why did they bother? "Likely" was the key word. Success was likely *most* of the time, but it seemed that females were willing to try and maneuver in the small space left to them. They tested the extent and the strength of their limitations at every opportunity.

Females were certainly as smart as males and as socially aware. Peggy and others proved this to me. Yet did they, surrounded by family, really need this smartness as much as the males did? Males had plenty of opportunities and a great need to exercise their social skills; how many chances did females have?

----

Exactly how smart were the baboons? The evidence was overwhelming that their intelligence included dimensions far beyond what anyone had imagined such "lowly" creatures capable of. Yet, reviewing the evidence, I was also struck by their limitations, emotional and physical.

Telling examples were to be found in the predatory tradition. It was clear that greater cooperation among males in a hunt would produce greater success in capturing prey. One avenue to stimulate cooperation might be to share the spoils between cooperating partners. But it wasn't so simple for the Pumphouse males, who seemed limited in both their willingness to either cooperate with one another or to share. The cooperation that did occur was restricted to alliances during aggression, usually with males of similar tenure in the troop against males of a different residency class. The rare, special mutual-assistance pact between two males paled by comparison with what families did for one another or what males did for their female friends. There was something in the evolutionary history of male baboons that incorporated the rule "Help your fellow male sometimes, but not too much."

Besides temperament, anatomy threw up obstacles to further predatory development. Since the baboons were physically unable to deliver a killing blow, the only prey they could tackle had to be of a size that could be restrained and eaten while alive. This meant that all the slightly larger tommy juveniles and even all but the youngest baby impala were off limits. I wondered if male baboon canines could become offensive rather than defensive weapons. Could they be used to kill a larger prey? I had my doubts, since their teeth were thin, long and relatively fragile. The older males in Pumphouse all had broken canines, and a few unlucky young baboons had already damaged theirs. This wasn't of much importance in terms of their aggression with one another; an old male without canines could still win a fight. But if male canines were to be the key element in a predatory attempt on larger prey, fragility *would* matter. Baboons had little else that would help them extend the range and type of prey they could handle. Of course the early hominids were in a much worse position—they lacked even the impressive baboon canines—but they solved the problem by inventing tools for killing and dismembering a wide assortment of species.

Females were limited, too, in their own ways: they were tied to the troop, emotionally conservative even in the face of such new opportunities as hunting and in finding possible loopholes in the dominance system. Why didn't Peggy attempt to regain her rank? Why didn't Thea? In most cases, females didn't question the switch in rank once a reversal had taken place. Adolescent males needed only one win to make their point with adult females. Even though it was inevitable that adolescents would become large enough to dominate by size alone, why didn't females snatch back their rank when they could, as Olive had once tried to do? There was something just as puzzling about the emotions and psychology of female baboons as there was about males.

Pumphouse baboons were smarter than anyone had ever imagined they could be. This was my most exciting discovery, doubly exciting because I knew I had only scratched the surface of baboon social intelligence. Yet for all their sophistication and resourcefulness, the world put up obstacles they couldn't surmount. At times the greatest obstacles were within themselves. Was it really different for any species?

# II. Implications

When I looked at the baboon world through "baboon spectacles," I saw a complicated landscape, characterized by sophistication and social intelligence, populated by animals with
long memories who, relying on social reciprocity, were of necessity
"nice" to one another. In this world, males and females shared a complementary importance in the life of the group. When I looked at the
baboon world through my own academic lenses, I saw what I had been
taught to see, which was something quite different. Many issues, including aggression, dominance and sexual roles, went out of focus as
I changed from one pair of glasses to the other.

It was now becoming obvious that the clear, sturdy, unambiguous
framework I had brought with me from Berkeley, simply buttressed
and functionally elegant, had become transmogrified. In its place had
risen a heretical, complex, slightly threatening structure, parts of which
were totally unexpected despite being well grounded in the day-to-day

realities of baboon lives. This structure grew as my discoveries grew, continually open to new findings. The foundations of my inherited framework—the concepts of aggression and dominance—had simultaneously become less important and more interesting when placed within the new view I was constructing.

Up until the 1930s, scientists thought of aggression as both abnormal and dysfunctional, since, to them, it appeared disruptive to the basic fabric of social life. The pioneering ethologists, including Konrad Lorenz, famous for his popular as well as scientific books, fundamentally changed this position. They began to look at animals from a new evolutionary perspective, one that inevitably turned aggression into an adaptive behavior. It became normal rather than abnormal, central rather than perverse; aggression was the way in which animals solved the critical problems of competition and defense.

As such, aggression came to be known as an important evolutionary feature of animal society. It soon also became vital in another way: aggression frequently resulted in dominance hierarchies which controlled and organized the interaction between individuals and, through those individuals, the group itself. Assuming, as most biologists did, that all essential resources were limited, an individual depended upon aggression and dominance for both survival and success. Thus, competition, reproduction, defense, aggression and dominance became inextricably linked in models of both animal and human societies.

Anthropologists such as Washburn contributed another dimension to the developing perspective on aggression: the importance of functional anatomy and a knowledge of primate evolution. Many physical differences between male and female monkeys and apes seemed to result directly from their different aggressive behavior. The conclusion was obvious: primates have the biological basis for aggression, use this equipment frequently and are highly rewarded when successful.

With this perspective in mind, Washburn turned his attention to humans, attempting to chart the development and possible transformation of aggression during evolution. First, it was obvious that humans lacked the nonhuman primate anatomy of aggression. Since aggression was an important means of communication about competition and defense, Washburn inferred that another factor must have been used instead. Human language was his prime candidate. If hominids could talk about issues of competition and defense, they might no longer need the special physical means to convey aggressive intentions, threats and displays. If so, language could have opened the way for a social system in which aggressive behavior was not constantly rewarded, as it appeared to be in nonhuman primate groups.

It seemed to Washburn that in the course of human evolution, language made possible a new and complex social life, a life that in itself modified the human body, emotions and brain. In fact, the specific part of the brain that makes language possible could really be considered the "social brain," to use Washburn's terminology, functioning as a mediator of social pressures and helping to produce appropriate social actions. Washburn and countless others, scientists and laymen alike, felt that aggression was deeply rooted in primate anatomy and physiology; it had had a long and important evolutionary history. But Washburn went further; he argued that during primate evolution, unique human events significantly altered both the biology of aggression and its role in the lives of individuals.

It was here that I began to depart from Washburn's ideas. What if aggression turned out to be neither so inevitable nor so central, not just among humans but perhaps among all the higher primates? What if the earlier position on aggression that marked a major advance in the study of animal behavior inadvertently and unconsciously *overemphasized* the role of aggression in the daily lives of many animals?

Even without my baboon evidence it would seem reasonable to expect that alternatives to aggression must exist. Any act of aggression held high risks: serious injury, even death. Wouldn't animals seek safer ways of achieving their goals? Even the strong and powerful might find themselves temporarily incapacitated. How would they behave then? What about the small and weak? Wouldn't they still have the same problems to solve? How could they do so without any aggressive advantage?

I was reminded of the relationship between predators and prey—the better the predators' strategies, the more pressure was put on the prey to find ways to avoid being caught. Shouldn't the same evolutionary processes that developed aggressive capabilities also have developed alternatives to aggression, as potential losers sought a way to not always be a loser?

The conclusion seems apparent. If aggression existed, then alternatives should also exist. Yet not all species might be capable of finding such alternatives. Social strategies among the baboons required that individuals assess complex situations, then modify what they might have done, based on past and present experience. This needs a special type of intelligence, an extremely retentive memory and enough brainpower to integrate many different pieces of information. Could ants employ social strategies? I had my doubts.

Viewed this way, it seemed obvious that primates are among the prime candidates capable of finding and using alternatives to aggression. Their brains and learning skills form part of the basic primate adaptation.

Where did this leave us?

Watching the baboons convinced me to take a new stance on the place and importance of aggression in the lives of animals. Aggression might be just *one* option instead of the *only* option that an individual could choose when he needed to defend himself or to compete with others. Furthermore, with alternatives possible, aggression suddenly would become less inevitable. Individuals should have acquired flexibility in their responses, being prepared for the possibility of aggression but not necessarily locked into reacting in an aggressive way.

Also, it seemed logical that the potency of aggression would decline once alternatives existed. Baboon social strategies would allow aggression to be circumvented, reversed or redressed. Was this why aggressive solutions apparently had been displaced from the lives of the Pumphouse Gang? What does an aggressive winner do when the loser gets around him without using aggression? In the end he is forced into a type of chess game, with checkmate achieved only when he counters with his own social options. When a male baboon grabbed an infant as a buffer against his aggressive opponent, the opponent had few choices. He had to grab another infant or a female, or else leave. Those who had no infant to use set themselves the task of making friends with females and infants, ensuring that they would have social alternatives in the future.

I was not suggesting that aggression was evolutionarily insignificant for baboons, for nonhuman primates or even for humans. Among monkeys and apes it was undeniable; many aspects of anatomy were linked to their aggressive behavior. Many differences between male and female bodies, where they existed, reflected their varying aggressive proclivities: the male's big canines, greater male muscle mass, shoulder mantles of hair and other bodily features used in aggressive displays. Yet no matter how ancient aggression might be, the social alternatives we observe in humans today—the ones that Washburn credited to the advent of human language—most certainly predate the hominids.

If we recognize that social tactics and individual flexibility in aggressive responses began much earlier in primate evolution than originally was believed (the thrust of my baboon argument), then it becomes more difficult to claim, as many had, that aggression is inevitable, not merely among humans, but among all the higher primates. For that matter, the argument extends to any other animal capable of creating social strategies.

What if inside us humans lurked the legacy not of a killer ape, but of a polite and sophisticated baboon? Would sports be a necessary safety valve for our aggressive *instincts*? Would armies be the inevitable result of an inability to contain our deeply rooted human aggressiveness, as some have claimed? If anything, as Washburn suggested, language would certainly improve social communication and increase social options for humans, further undercutting the necessity for aggression in daily life.

―――――――

Without the central cornerstone of aggression in animal society, could dominance still play its crucial evolutionary role in structuring individual lives and animal groups? Certainly the evidence from the Pumphouse Gang suggested that dominance played a minor role among males and an as yet enigmatic but pervasive role in the lives of females and their families. Yet families and friendships were at least as important as hierarchy in providing organization and stability to the female core of the troop. Neither the baboon evidence nor my new argument could be stretched in any way that would produce the male dominated, male hierarchically organized society of the old baboon and baboon-based human models. How important was dominance and hierarchy to society, whether baboon or human? I would have to search further for the answer to that question, but I knew that the old answers were no longer satisfactory.

―――――――

The original views of aggression and dominance also carried with them suggestions about the evolution of differences between males and females and the "natural" division of roles between the sexes in human society.

A society based on aggression and male dominance (the baboon model) was augmented further along those lines when early hominids took up a hunting way of life, this evolutionary argument claimed. Women now stayed at home, caring for children and gathering nearby plant foods while men went out to hunt. Women were unable to hunt because of several handicaps: they were always encumbered with children and couldn't run as well as men because of changes in human anatomy. The trend to a larger brain during human evolution (culminating in a brain that is more than triple the size of ape brains) came into direct conflict with another anatomical trend. In order to walk efficiently on two legs—bipedalism is a hallmark of human adaptation—the pelvic diameter had to be as small as possible. Larger-brained babies needed exactly the opposite, a large pelvis for safe passage through the birth canal. Human females were saddled with a compromise: their pelvises were larger than those of the males, giving critically needed space to babies during birth but making them less bipedally efficient, particularly when running.

These handicaps were compounded by a change in the condition of human newborns. As brain size continued its remarkable expansion, it was also likely that human infants born at a more premature point, by monkey and ape standards, survived the hazardous birthing process better than full-term babies. The more immature human newborns became, the more dependent on maternal care they were. They lost even the most rudimentary monkey and ape abilities, such as being able to cling to their mothers at birth or being able to move around on their own shortly thereafter.

Arguments about the "natural" division of roles between males and females in hominid groups were based on this scenario. Females were inevitably confined close to home by domestic responsibilities, while males were forced to meet the new challenges of hunting. Taking such risks also held advantages; hunting activities conferred a special status and power on males. The division of labor and roles was completed by an exchange of goods between men and women: meat for vegetables, and perhaps, as some suggested, sex for meat when males bought favors from females or females used sex as a special resource to barter for food and protection critical to their survival and that of their family. Because hunting was a way of life for humans for more than 90 percent of human evolutionary history, the hunting society, its roles, underlying emotions and psychology were deeply embedded in human biology and difficult to erase from later human behavior.

In the last ten years, a groundswell among the feminists has tried to change what is viewed as a "sexist" model for the evolution of human society. The debate rages on, as proponents of female rights and female power try to demonstrate that gathering was more important than hunting to our ancestors, and that consequently women were more important than men in hominid hunter-gatherer communities.

What I had observed among the Pumphouse baboons strongly challenged some of the traditional arguments about human sex roles and their evolutionary origins. Unfortunately for the feminist position, the baboon evidence did not support the reversal of roles they sought. Instead, the baboons suggested that there was an array of sex differences in behavior, emotions and psychology in the primate pattern that predated the innovations of hunting. Yet there was equally strong evidence for a complementary equality of male and female in most social domains, including politics and caretaking. More provocative still was the discovery of social reciprocity among individuals. This meant that even without bipedalism, without the invention of tools, without hunting and gathering, without the development of the large human brain, with its capacity for language and culture, and without extremely dependent infants to restrict female movement, male and female baboons are involved in a complicated exchange of favors.

Such baboon insights are both provocative and complex. Male and female psychologies are already very different among baboons. There is no doubting that females are conservative in trying new behaviors like hunting or in breaking out of old patterns like family rankings; they are unwilling to take risks or strike out on their own. They prefer the safety and reassurance of being with the group and staying close to family and friends. Stability is a goal that they achieve a remarkable amount of the time, sometimes in the face of formidable odds. Males are just the opposite: dynamic risk takers, they seize as many opportunities as they can. If stability and a status quo are indeed their goals, these are elusive, at least in male relationships with other males.

To me the baboon facts suggested that the division of roles others credit to the new hunting life-style of the hominids was already present among the Pumphouse baboons. Their relatively minor predatory adventures illustrated but did not create the differences between male and female proclivities. Having meat to share, as among hominid hunters, might increase the variety of exchanges between males and females, but neither reciprocity nor sexual politics originated with hunting and gathering. Although baboons shared only meat and no other foods, male and female friends bartered social and sexual favors in a complicated tit for tat that formed the foundation of their friendship. Males also bartered favors with their infant friends, and to a much lesser degree, with their male allies.

The final point that the baboons made is perhaps the most striking: there is nothing inherently powerless or impotent in female roles, no matter how involved these are with raising young and being the stable core of the group. The old arguments assumed that these were the only roles and functions females had: but among female baboons, they formed the basis for a variety of other political and leadership positions. When knowledge and social influence create power, as they do among baboons (and among humans), females are socially powerful because they have both. More important still, whenever alternatives to aggression exist, even giants find themselves needing the smaller or weaker members of their society in order to create effective social options for their own survival. Experience, social skill and social manipulations were important to individual baboon survival and success. Real power resided with those who were "wise" rather than those who were "strong," those who could mobilize allies rather than those who try to push through with brute force.

Underneath all the cultural posturings and institutional edifices, I don't think that modern humans are any different from baboons in this regard. Those who believe that we are different will have to find new evolutionary explanations for how humans changed.

--------

As I documented the importance of social strategies to the Pumphouse baboons and the new dimensions of complementarity and equality that these strategies introduced into baboon society, I found myself faced with another issue. Were social options and social sophistication necessary or possible to the same extent for both sexes? Was the evolutionary path identical for males and females?

From a male's point of view, I could see why social strategies were necessary. First, competing males were potentially fierce and dangerous opponents. Males could inflict real damage on one another, even though they exhibited incredible self-control most of the time. Restraint inevitably breaks down. Social strategies gave males effective ways to counter the risks of real aggression.

A second factor at work was that the males left home. In that first move they broke all the ties of kinship. From that moment on, a male could no longer rely on his family's help. An older brother helps out for the very reason that he *is* an older brother, not because he expects to be helped in return. The same is true of parents and children, aunts, uncles, nieces and nephews. Given the peculiar age structure that can develop within baboon families, even a cousin could sometimes be of major assistance.

Once a male left home, he had to create his own allies should he need them, and the males joining Pumphouse did so. By investing time and energy in other troop members, they created "families," individuals related to them by ties of social reciprocity. These relationships produced the necessary allies when they were needed.

Social sophistication, social finesse and social manipulation served a male well. They had to, because unless he wanted to go it alone, they were his only choices. Females faced a completely different situation. They were always surrounded by their relatives and could get help in a second should they need it—or at least after a loud scream. Of course, such assistance worked both ways, but unlike the social reciprocity of friendship, no one appeared to be keeping score. "I helped Aunt Marcia three times last week and she didn't help me at all, so I won't help this time" didn't seem to operate among baboon family members. The family system was tight and effective. Occasionally a female might try to break out, enlisting those outside her family in order to better her own position, but such methods rarely worked.

Given the constraints and opportunities of family relationships, females would probably run into fewer situations requiring the social sophistication that was the basis of male survival and success. Yet females were smaller and lacked the aggressive anatomy and potential of males. They sometimes needed the help of friends and social sophistication to get what they wanted, as Peggy's behavior with Dr. Bob's tommy carasses illustrated.

So while males and females both require social strategies, it seems likely that they emphasize different tactics: males rely on systems of social reciprocity they can actively construct, while females can usually depend on assistance from family that comes almost automatically.

I already had good indirect evidence about male and female differences in temperament, emotions and psychology. Could the difference between the social tactics of the sexes have biological consequences as well?

When males invest in others socially, and later manipulate them, they constantly have to test and assess, to evaluate costs and benefits with a considerable degree of sophistication before, during and after each of their social interactions. By contrast, females may have less need of these skills since they rely primarily on family, who will help under most circumstances.

Social strategies depend on just those mental faculties that we know were involved in the expansion of the brain during the course of primate evolution: memory, integrative capacity and the ability to weigh complex situations. Although some abilities would be important to both male and female social strategies, the possibility remains that the differences between males and females in the amount of social sophistication they need could have had an impact on brain structure.

The general primate trend was accelerated during human evolution. We know that differences exist in humans in the specialization of the two sides of the brain, and there is also controversial data showing sex differences in spatial and verbal abilities. These differences are what we would expect if social strategies had an influential role in brain organization during human evolution.

---

It seemed to me that the study of baboons and other monkeys and apes was even more important in reconstructing human evolution than the earlier simple models implied. Baboons could help to provide not just examples but important principles. As we better understand how evolutionary changes occur in complex primate groups, we also come to understand what is possible for primates. This is the best way to set the starting point for the scenario of human evolution. We are left to discover what was probable for the early hominids by placing what is possible within the context of hominid biology, anatomy and ecology. In future, evolutionary reconstructions will thrive only if they focus on processes and governing principles in primate societies rather than on single species, no matter how similar we think those species are to us humans.

Although I was committed to this new kind of approach, and did not want to reinstate a baboon model of human evolution, I was constantly struck by how much more like humans the baboons now seemed. They learned through insight and observation, passing new behaviors from one to another both within a single lifetime and across many lifetimes. This is social tradition, the beginnings of what eventually became "culture." Their social maneuvers and social sophistication raised them to new heights. The cumulative effect of my years of watching, analyzing and interpreting baboon behavior was that I felt more and more at ease in applying human terms and concepts to the baboons.

I had no stake in human uniqueness: each species is unique. But the similarities intrigued me. It was the similarities that made it possible to imagine how behaviors could evolve through time. How would hominid innovations occur? It was hard to believe in sudden large leaps into thin air without some preexisting material to work from. Evolutionary reconstructions of human behavior become more believable the greater the continuum between human and nonhuman primate behavior. The baboons managed to add many new stepping-stones to that route.

Talking about baboons in more human terms left me open to criticism—that I was projecting human behaviors onto these nonhuman creatures which they did not, in fact, possess. And I wanted to go further still: if human terms were appropriate to monkey *behavior*, wasn't it possible that baboons might be considerably more humanlike emotionally, psychologically, intellectually? My university training had profoundly impressed on me the dangers of anthropomorphism. Yet it was the baboons themselves who were responsible for my change of heart. I was a reluctant voyager; I had resisted them at every step, yet now, having traversed the entire road, many of the old distinctions between baboons and humans blurred.

I turned to history to see if I could find any support for this change in my position, and as I reviewed the development of ideas about animals during the last few centuries I found some surprising but reassuring answers.

The same animals could be saints or demons, stupid brutes or sentient beings, depending on the culture and the era. A remarkable change in thinking was brought about as a result of Darwin's ideas. Westerners, who at that time believed that nonhuman animals—monkeys and apes in particular—were considerably lower than the angels, were now faced with a conundrum. If humans were truly related to these creatures, then either we were baser than we thought, or else— and this was much more appealing—*they* must be nicer, brighter and more humanlike than we had first painted them.

The post-Darwin period saw animals take on a sentimental appeal; animal abilities were elevated to a degree that we, as humans, could be proud of. The *animals* hadn't changed; it was *human* expectations and interpretations that had turned around. This was why the ethologists of the 1930s became so adamant against attributing human qualities and abilities to animals. They recognized how inappropriately it had been done, and instead wanted to be able to understand animals in their own terms.

But just as the pendulum had swung one way with the Victorians and post-Victorians, the early ethological approach made it swing equally drastically the other way. The ethologists urged that complex behaviors be reduced to the smallest possible unit and later reassembled into meaningful clusters. They recognized the importance of having rigorous methods and hard facts with which to back up statements and conclusions which, otherwise, could be fraught with biases and projections. The problem came in choosing just what to study. Important areas of animal behavior were difficult to study in this rigorous fashion. Emotions, psychology and the mind all fell by the wayside as the modern investigators began with what was easiest to tackle well.

Initially, scientists felt we would someday learn just how to study these important but difficult areas. Soon what had been omitted for the sake of convenience began to appear irrelevant. The position that it was hard to study these issues subtly shifted to the position that these were unimportant aspects of behavior, that we could explain what an animal was doing simply by referring to its outward behavior. Researchers who talked about what animals might "feel," "think" and "decide" were being unscientific and sentimental. The efforts of dedicated scientists had inadvertently robbed nonhuman animals of many of their important abilities.

The line between unwarranted anthropomorphism and what I came more and more to see as simply giving back to the animals abilities that they had lost to historical circumstances was a narrow and difficult one to walk. I was still uneasy with the idea that the longest-term resident males decided to leave Pumphouse because they *understood* that if they stayed they might enter into unsuitable matings with their own daughters. At the same time I could accept the idea that Peggy knew exactly what she was doing when she groomed Dr. Bob into a stupor and then stole the tommy carcass from him. I was convinced that she and Olive both acted premeditatively and manipulatively when Peggy attacked Dr. Bob's friend in the escalating competition over the meat he possessed, and when Olive screamed at Toby later during the day on which he had first dominated her.

I had many male examples as well. Males used sophisticated tactics around consorts, convincing others that they weren't interested while arranging to be at the right place at the right moment to claim the female. Males incited the aggression of other followers, ducking out just when the going got rough, only to appear again at the critical moment. Males harassed an unsuspecting infant on its way to be groomed by its mother and then whisked the female away from her confused consort in the pandemonium that ensued.

Were Pumphouse baboons conscious of what they were doing? It now seemed to me that they were no less—and perhaps much more— aware of their actions than most humans.

Was it anthropomorphic to believe that baboons were intelligent manipulators of their social world; to think that they weighed alternatives and made decisions; to think that even without language they had mental symbols that allowed them to think first and act later, that allowed them to create remarkable unwritten contracts of social reciprocity? Upon reflection it seemed peculiarly human, particularly anthropo*centric*, to *deny* them these abilities. As I searched further, I found that a few other investigators of animal behavior were reaching similar conclusions.

---

Working through my years of data, constructing new ideas, testing them with what I already had in hand and then with new evidence that I was able to collect finally convinced me that my baboon spectacles rather than my academic lenses were reliable. As heretical as some of my ideas might sound, they helped me to predict what a Pumphouse baboon actually would do much better than the previous theories. To me, that was the acid test.

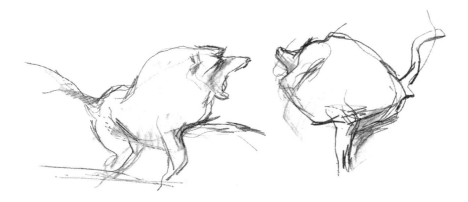

# 12. Woes

I was excited about my discoveries, but I was worried, too. When I began my investigation of the males in 1976, I suspected that I was venturing into academic waves of capsizing dimensions, but I had successfully delayed thinking about such problems. Now, in 1978, they had to be confronted.

The idea that there was a stable female core to a group of baboons, that males were transients and females took on many of the roles previously assigned only to males, was no longer heretical. There was a rapidly growing body of information on nonhuman primates that pointed in the same direction. Even the issue of culture and tradition as I had linked it to Pumphouse predatory behavior was not especially controversial.

Still, I knew my work painted a picture of baboon society that others would find difficult to accept. My shocking discovery was that males had no dominance hierarchy; that baboons possessed social strategies;

that finesse triumphed over force; that social skill and social reciprocity took precedence over aggression. This was the beginning of sexual politics, where males and females exchanged favors in return for other favors. It appeared that baboons had to work hard to create their social world, but the way in which they created it made them seem "nicer" than people. They needed one another in order to survive at the most basic level—the protection and advantages that group living offered the individual—and also at the most sophisticated level, one marked by social strategies of competition and defense. They also seemed "nice" because, unlike humans, no member of Pumphouse possessed the ability to control essential resources: each baboon got its own food, water and place in the shade, and took care of its own basic survival needs. Aggression could be used for coercion, but aggression was a roped tiger. Grooming, being close, social goodwill and cooperation were the only assets available for barter or to use as leverage over another baboon. And these were all aspects of "niceness," affiliation, not aggression. Baboons were "nice" to one another because such behavior was as critical to their survival as air to breathe and food to eat.

What I had discovered was a revolutionary new picture of baboon society. Revolutionary, in fact, for *any* animal society as yet described. The implications were breathtaking. I was arguing that aggression was not as pervasive or important an influence in evolution as had been thought, and that social strategies and social reciprocity were extremely important. If baboons possessed these, certainly the precursors of our early human ancestors must have had them as well.

Such statements were provocative and would be challenged. When the time came to defend myself, I felt poorly equipped. I was more than an academic interloper; I was now culturally alien as well, even among my colleagues.

Claude Lévi-Strauss wrote that being an anthropologist *means* being alienated, being someone for whom the world is divided into "home" and "out there," the domestic and the exotic, the urban academic environment and the tropics. First you travel as a stranger to another culture. After years spent with strangers, you return home, still a stranger. The anthropologist has made detachment a profession and can never feel "at home" again. Had I read Lévi-Strauss earlier in my career, I might have been better prepared for what the baboons were going to do to my world.

Although during the last few years, between 1976 and 1978, I divided my time between teaching in California and baboon watching in

Kenya, I was only truly happy when I was with the monkeys. They were my emotional center, and an important part of me remained with them even when I was physically distant. The California life-style that had spawned me seemed alien, compartmentalized, unintegrated. The feeling of wholeness that came from studying and living with baboons was gone. I looked at the photographs of myself with the monkeys wistfully; there I saw a happy person, someone at peace with herself and her environment. The image confronting me in the mirror in California was full of tension and the unhappiness of not being able to cope. Yet I had everything I wanted: field experience, a good university position and a budding career.

The problem was that the baboons had become a passion, almost an obsession. When I was away from them I surrounded myself with surrogates: photographs, tape recordings. I was at ease only when I could talk about my other life, my life with baboons, with the few people who listened and understood.

Later I learned that what I was feeling was not unusual. A study of the reentry problem of animal watchers suggested that they have even more difficulty than those who investigate foreign cultures. Cultural anthropologists run the risk of being alienated from their own societies through their experiences elsewhere, but at least they have been living in a world of people. Imagine being in a land where no one speaks at all, where contact is forbidden and yet incredible intimacy is the norm. Returning to the human realm is difficult; for some it becomes impossible.

Such was my frame of mind when I ventured into the scientific community with my ideas about baboons. This wasn't my first foray into the academic arena. I had first presented my data and interpretations of baboon predatory behavior in 1974, and the response had been enthusiastic. My facts were important, my speculations interesting and valued and criticism was light. But dominance, aggression and males were a completely different story.

When I told my nonacademic friends that I was arranging a meeting of world baboon experts, they regarded me with disbelief. How many could there be, and what would they have to say to one another that could take ten whole days? Even before I began studying Pumphouse in 1972, baboons were already the most studied nonhuman primate. Even more baboon studies were initiated in the following six years. By this time there were two study groups in Tanzania, three in Kenya, several in Ethiopia and a few in southern Africa, all devoting serious efforts to understanding baboons. In fact, it was difficult to decide which eighteen people I should invite to the meeting, eighteen being the limit that the sponsors, the Wenner-Gren Foundation in New York, always set for their internationally renowned conferences.

The final selection was a compromise. The group included someone from each major site, with a balance between senior scientists—or "silverbacks" as they are called among primate watchers—and quite junior researchers, some of whom hadn't even finished their doctorates but had spent at least a year watching baboons.

I had two aims. The first was simply to exchange ideas about baboons, integrating investigations, making comparisons—and perhaps in so doing discovering information that might otherwise escape our attention. I also had a more philosophical purpose. An anthropologist has to adopt a very ingenious stance about his own doubts and uncertainties. How was it that my data differed so much from that of everyone else? Why were my interpretations at odds with past and present views? Were the Pumphouse baboons really so different, or did my results reflect differences between my methods and how others had done their studies?

How did *any* of us reach conclusions? Did the way in which we chose the topics we would investigate influence what we would find? Why did some answers seem more satisfying than others? It was difficult for me to believe either of the two alternatives that confronted me: that I was wrong and everyone else was right, or that I was right and everyone else was wrong. I had returned to the baboons several times between 1974 and 1978, and each time left feeling more convinced that I now possessed some basic principles that allowed me to predict what they would do. But something was wrong, and only deep probing would uncover it and set it to rights.

I had not set myself or the conference members an easy task. Tension and dissension were already in the air; one contingent of silverbacks felt I was too junior to be organizing a conference of worldwide importance, and other juniors agreed with them. Trying to keep the temperature as low as possible, I aimed for a modest presentation: I would demonstrate simply and directly that male Pumphouse baboons did not have the traditional dominance hierarchy, while females did. Then I would discuss how in the past we had reasoned our way through the labyrinth of baboon behaviors to the conclusion that male dominance gave the troop its basic structure.

The worst ten days of my life followed. Since we had circulated all our papers among one another in advance, my conclusions were already known. As I was preparing to speak, I discovered that I did have enemies: an archrival sitting nearby was openly recalculating my statistics on a pocket calculator.

At the end of my presentation, no one spoke. The polite silence was finally broken with barely guarded accusations: I had invented my data, I didn't have enough information to draw the conclusions I had come to and that there *had* to be a male dominance hierarchy among Pumphouse males. I had managed to miss it, that was all.

Lunch followed. I was upstairs in my room; later I learned that everyone assiduously avoided the topic of the male dominance hierarchy. Had there at least been a discussion of the subject? Did no one believe me, or see the value of what I was trying to do? Was the idea of male dominance so deeply rooted that it couldn't be challenged?

The afternoon session provided no answers. My paper was followed by one that argued for male dominance hierarchies among three troops of baboons. Accolades circulated around the table afterward. I was stunned. Here was a paper with a fraction of the data I had, and the "proof" everyone accepted was a computerized simulation of the missing information! If *I* had too little data, how could my colleagues accept a conclusion based on so much less? Were baboons assumed to have male dominance hierarchies until proven otherwise? What *was* necessary to convince my colleagues?

The conference was a turning point for me, intellectually and emotionally. My only allies were two senior researchers, one of whom was surprised at all the excitement, since for him male dominance was a dead issue. The other both empathized and sympathized; she had tackled male dominance a decade earlier and received much the same reaction. Many participants were unwilling or unable to discuss the major philosophical issues on our agenda. Almost no one wanted to probe into their reasons for choosing the path they had taken, to explain why they had found some questions important and some trivial, or why they accepted some answers and not others when the data was equally compelling.

Yet amid all the disappointment I did find a kindred spirit. I had invited Bruno Latour, a French sociologist and philosopher of science, to the conference, even though I had never met him. He had written a remarkable study of how scientists do science. I wanted him to be our conscience, to help us explore the more difficult issues. His reception was chilly; there was even a move to evict him. I admired his calmness during and after the attack.

I left the conference with new insights, not about baboons but about baboon watchers. Bruno helped me realize that I would have to be more politically astute if I wanted to gain an audience for my ideas. Instead of claiming that there was no male dominance hierarchy, I decided I would have to uncover one, no matter how insignificant it was to the baboons. Perhaps then my colleagues would listen to what else I had to say.

Mozart, Bruno reminded me, advised: "Happy is the man who loves reason and yet who accepts the world as it is." Many of my colleagues were "technicians," narrowly working through the fine details of traditional ideas without questioning or doubting. They were not interested in understanding why or how they did what they did, perhaps even were threatened by the idea of reflecting on it, philosophically and historically.

I recalled a stunning Washburn lecture. "A 'fact' is only the current best fit between reality and our methods to study it," he had said. "As our methods change, so will the 'facts.' " That was a far cry from the notion of science as the "truth," the objective and precise apprehension of reality. But my experience made me believe Washburn's position. And Bruno helped clarify what I had previously suspected: social, cultural and even psychological factors intrude on science because scientists are human. That doesn't diminish the endeavor, but it does change it. With these obstacles to overcome, good science was a much greater achievement. We vowed to pursue the matter. Bruno wanted to make his next study a closer investigation of baboon watchers!

––––––––

As early as 1973, long before that terrible conference, I had tried out my ideas on visitors to Pumphouse and colleagues in Kenya and elsewhere. Their responses damped my enthusiasm. Some of the reactions were just silly. Rival baboon watchers criticized me when I decided to leave the van and do my study on foot. I was "interfering," ruining the baboons! Another time, when I reeled off the names and sizes of Pumphouse families, they said I must be wrong—"baboons can't have families that large." My descriptions of friendships between males and infants and males and females were greeted with disdain, despite Tim Ransom's earlier accounts of just such friendships. And my conclusion that males didn't really have a meaningful dominance hierarchy met with downright hostility.

I felt both smug satisfaction and anger when these same detractors later discovered the joys of moving among *their* animals, or found for themselves what they should have known all along—that baboon family size reflects the survival rate of infants, which will vary at different times and in different circumstances. But what was I to think when they announced an exciting new discovery—that baby baboons had "godfathers"—males who befriended and protected them?

More than mere silliness was at work, and more than just hurt pride. I was naïve; I had imagined that one did the research, gathered the information, analyzed, interpreted and presented it to the scientific world. Then the work would be evaluated and incorporated, if accepted, into the basic knowledge within the field. But there are cliques in science as in any other facet of human endeavor. If you are part of the "in" group, even minor findings are discussed and integrated, eventually becoming part of the working knowledge in the field. If you are not part of the clique, you stand a good chance of being ignored. Except for my work on predation, which wasn't particularly controversial, no one replied to or cited my work on dominance, aggression, social strategies and friendships. I was prepared to be wrong or to modify my position in the face of cogent arguments from colleagues; I was not prepared to be nonexistent. Only a few other heretics gave me their reactions. They liked my ideas, but warned me that people would have a hard time accepting what I was proposing.

As time passed the tide began to turn. Social sophistication was being uncovered everywhere, although the ideas about it weren't presented in the way I had suggested. Aggression was beginning to loosen its powerful grip over interpretations of animal and primate behavior. Should I be satisfied with having pointed out the new direction much earlier, even if no one had listened? Was science about discovery or about influence? My mind said "discovery," but my guts answered "influence."

It was all too easy to get caught up in emotional, heart-wrenching pettiness; science is no different from any other human occupation in that respect. I laughed when I read an editorial in *Science* written by a man who had left the higher echelons of the business world for the quieter fields of academe, hoping to rid his life of the cutthroat politics he had found there. His article was announcing his return to the world of finance. He couldn't take the competetiveness, underhandedness and frenetic pace he found among scientists. The business world was mere child's play by comparison. I was beginning to understand.

It was at this time that I discovered a remarkable book about how science "works," *The Structure of Scientific Revolution,* written in 1962 by Thomas S. Kuhn. It seemed to echo the ideas of both Washburn and Bruno Latour. Kuhn rejected the traditional view that there *is* truth out there and that science simply uses special techniques to apprehend it. His main concern was with how ideas change and what produces a scientific revolution; his observations agreed with both my own personal experience and with what I knew about the short and relatively modern history of the science of studying primates.

In science, according to Kuhn, ideas do not change simply because new facts win out over outmoded ones. Many more social, cultural and historical variables make up the complete picture. Since the facts can't speak for themselves, it is their human advocates who win or lose the day. My friend Bruno Latour had documented just this in *Laboratory Life,* his 1979 study of a laboratory whose director had just won the Nobel Prize. But Kuhn, studying the "hard" sciences, went further. Science operates by accepting models (paradigms) of how the world works. Once a paradigm is in place, it is very difficult to dislodge it, and when discrepancies are found, they are either ignored or written off as the results of bad methods or "bad" science. But these discrepancies eventually become overwhelming, and another version, model or paradigm replaces the old one. This is what is known as a scientific revolution. After the revolution, the process doesn't change greatly. Most scientists devote their time to "normal" science, to use Kuhn's term—working out the fine details of the new model instead of the old—and are reluctant again to admit divergent ideas or information.

Kuhn calls this a "gestalt switch." Once there is a new model, scientists can look in the same places they have looked for decades and make new discoveries. They can look at old bits of information that once seemed irrelevant or trivial and find important new "facts." Equally fascinating is Kuhn's observation that revolutions are often initiated by an outsider—someone not locked into the current model, which hampers vision almost as much as blinders would. The advantage of actually being within a certain paradigm is that great progress can be made in understanding in detail and in depth the world as it is revealed by that model; the difficulty is that each model narrows a scientist's vision, and paradigms can have great power over how people think.

Furthermore, as Kuhn's research proved, humans tend to be conservative. It is often difficult for a scientist to switch paradigms—in the course of a lifetime to change his or her ideas. Only when the adherents of the old model actually die does the new one come fully into its own.

This theory seemed to fit my understanding of primate research, although some scientists—those involved in nuclear physics, for example—might put up an argument about whether research into the behavior and evolution of primates is really a science at all. This partly explained why the baboon male dominance idea had held sway for so long. I had uncovered many examples of the power of a model or idea to mold thinking. A good illustration was C. R. Carpenter's work on howler monkeys, the New World monkeys he had studied in the 1930s. In his original description, these primates were almost communists. There were no hierarchical relationships among males or females, no sexual competition; daily life was marked by communality. Yet in a paper Carpenter wrote thirty years later in the heyday of the baboon model of primate society, a model based on male competition and dominance hierarchy—a paper that contained no additional data on social behavior—he transformed howler monkeys from noncompetitive socialists to members of a society where males possessed dominance ranks and competed with one another for females! This transformation was not conscious, but it was telling evidence of how powerful a certain view of the world can be, even among reputable and rigorous investigators.

What I was seeing in Pumphouse society, and what was now being increasingly observed in other studies, was contributing to the beginning of a paradigm shift. I could see the discrepancies building up, and it was becoming more and more difficult for others in the field to ignore them, to simply write them off as bad methods or bad science, as I had been accused of at the baboon conference.

Nevertheless, there was more resistance to certain ideas in primate research than seemed warranted. Bruno Latour and I decided to explore the issue by combining my anthropological background and his knowledge of philosophy, sociology and history of science. We made a selection of authors who had written about the beginnings of human society, including with the most famous those who were more obscure: some names everyone would recognize and some who had only a small following—Rousseau, Hobbes, Nietzsche, Freud, Aristotle, Engels, Bentham, Mauss, E. O. Wilson, Richard Dawkins, Robert Trivers,

W.D. Hamilton, Richard Leakey. Then we began studying the logic of their arguments and the role that the available facts played. The comparison proved illuminating.

The logic of the seventeenth- to nineteenth-century philosophical accounts was vastly superior to the logic of the modern scientific accounts. The explanations of Rousseau and Hobbes held together and were more logically compelling, even though these writers had no or few facts to prove that human society began in the way they were suggesting. They were not saying *This is the way it was* but *This is the way it should have been.* The modern accounts, while bolstered by scientific theory and full of contemporary information about biology, primates and fossils, seemed to state the basic facts and then fly off into the realm of fantasy. Facts and conclusions were related only by their simultaneous appearance in the same book or article; there were no logical links, scientific or otherwise.

Bruno and I had had no preconceived ideas about what we would find, except that we anticipated that the modern scientific accounts would be less like fables and stories about our human origins than the older, clearly philosophical works. Facts would play an important role in modern accounts but none in the older ones, since these facts were not known then. Yet "anything goes as long as it is in English" was Bruno's conclusion about the modern scientific authors; arranging "facts" in a logical and coherent way was of secondary importance.

Why should this be? We could only guess. Discussions of human social origins, whether labeled scientific or not, take place within the broader context of the needs and expectations of a living society. This creates special difficulties. Why do some versions of our origins seem more satisfactory and pleasing than others? Why do we favor one over another? There are hidden preferences, which play an important role in our choices.

Bruno and I agreed that most humans are constantly searching for an understanding of themselves, of who they are, where they came from and how they are expected to act. Every time someone tries to account for the origins of human society, another genealogy of human society is created. Expectations are set up about what is natural and what is not, what is justified and what is not. This information on our "origins" is used to explain the facts of modern life and to justify a specific set of conditions—moral, political and economic. We recognize that this is true of popular uses of scientific accounts, but it also seems to be true of the scientific accounts themselves. Bruno reminded me that it is important not to forget that scientists are humans too, needing justifications and rationalizations for living their life the way they do, just like everyone else.

One could say that the scientific accounts of human social origins are the functional equivalent of myths. This need not be a pejorative statement. In primitive societies as well as in modern industrialized nations, myths are created in order to handle such timeless problems as self-definition and allegiance, the relationship of humans to animals, and the purpose of living in society. Neither Bruno nor I felt that we should return to prescientific answers to these questions, but our work together did help to explain the tenacity of certain modern "scientific" ideas. It also warned us that perhaps our scientific accounts were not as scientific as they should have been, especially when dealing with certain topics. In the future it will be important to utilize facts more rigorously, while at the same time being aware that the entire topic of human behavior, its origins and past evolution—and this includes closely related animals—is highly charged emotionally.

Science, it was clear, was complicated because its practitioners were human primates. Primates were not created as objective machines. Every primate, humans included, sees the world in a particular way— through primate eyes, ears, noses and hands, but, most importantly, through a primate brain that has a very special history, a brain that is extremely skillful at performing certain functions and extremely poor at performing others. It is a unique attribute of humans that at least some of them want to try and understand the world "as it really is" rather than being satisfied with seeing and comprehending it only to the extent that it serves some immediate and selfish purpose.

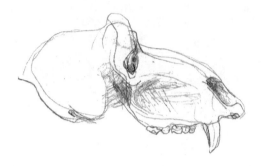

# 13. Crop Raiding

It was with great relief that I turned from the world of academe to the world of baboons, only to find that it, too, was being severely threatened.

In 1976, just as I had begun my study of male adolescent baboons, the Coles had sold Kekopey. I had been devastated. Hugh Cole, the middle son, had been like a brother to me. His grandmother's china coffeepot, which had been at Kekopey for fifty years and which he had given me, is one of my treasures and is still a potent reminder of the past.

I was never certain why the Coles sold Kekopey. Rumors were plentiful. The official story was that Arthur's failing health had made it necessary for them to return to England, but the rumors also hinted that political pressure was being applied, since Kekopey was an extremely desirable piece of land in an era of increasing Africanization.

There had never been any official arrangements between the Coles and any university or research agency. Each year of baboon research was made possible by the kindness of the Coles, and was negotiated directly with them by those involved in the research: first Bob Harding, then Matt Williams, then—until the Coles left in 1976—me. After they left, I negotiated with each new ranch manager, with the agricultural cooperative that bought Kekopey and eventually with the individual owners of numerous five-acre farms.

What would happen to Kekopey now? The plan was to move in landless people from other areas, dividing the ranch into small farms or *shambas.* Whether these would directly invade or overlap Pumphouse's range or whether they would be peripheral remained uncertain.

Ironically, I had joined forces with Bob Harding only shortly before the ranch was sold. Observations had limped along from one person to another for six years, and it now seemed time to put the project on a more formal basis, thereby guaranteeing that there would not be a break in baboon watching and that permanent project records could be kept. The hospitality of the land and the landowners, along with the exciting results of the baboon studies and the important role these insights can play in our interpretations of our own evolution, had convinced us to make the study of baboon society a long-term effort. The Gilgil Baboon Project was officially born in 1976. The baboons were watched full-time by many United States and European university researchers. I was the project's director, along with Bob Harding, and the baboons' most frequent observer. I was delighted by all this. I would have to be in Gilgil constantly, supervising the graduate students, and would even be able to keep in touch when I was back in California. I knew the baboons would never write to me, but the graduate students would have to.

But now the sale of Kekopey meant that the Gilgil Baboon Project, its long-term records and the stream of graduate student researchers would become pipe dreams. We were all temporarily evicted from the Red House while surveyors began to prepare the 45,000 acres, dividing them into 5- and 10-acre parcels of land. Then, just as suddenly, we were all allowed to return. My heart sank as I watched the surveyors place sticks everywhere, marking off plots that were to be purchased by the shareholders of the cooperative. I sat in numbed silence as truckloads of people drove around the baboons' home range to locate their piece of land. I had nothing against the fifty needy widows who were being given a special deal—or against anyone else for that matter. I simply didn't want people to settle. To my mind they were intruders who had no right to be there; I was on the side of the baboons.

Nothing happened for what seemed a long time. A few farmers arrived. Some houses were built near one of the baboons' favorite sleeping sites. Curious monkeys investigated the new "playthings" so generously erected for them. All I could do was clean up after them when they were tired of playing. The mess was usually fairly innocuous and almost unnoticeable: empty Coke bottles in new locations, rough-cut timbers covered with a new aging chemical, baboon urine. Once the baboons did manage to get inside a hut and emerged with a large ball of yellow plastic twine. A juvenile took the lead, unwinding it as he went round and round the hut. By the time he was finished, only a large bow was lacking to make the hut look as if it had been wrapped for a birthday present.

These houses didn't last long. The farmers were mysteriously relocated into areas where the baboons seldom ventured. Disaster was averted, and I returned to California reassured, with more data and the idea of seeking larger grants that would permit me to watch the baboons far into the future. I wanted to say forever, but no one funded forever. Although three years passed without any major problems, I still could not rid myself of an ominous sense of impending doom. I was glad, however, that I hadn't given up the baboon project in 1976, as so many had suggested I do. Three years later, armed with considerably more data, the future didn't seem in jeopardy. I continued to stand ready as the baboons' protector.

———————

In the summer of 1979 I returned for a teaching stint in California feeling smug and satisfied; the Gilgil Baboon Project was doing well. Within three weeks, entirely without warning, everything changed. Pumphouse began to raid the crops so carefully nurtured by the farmers. A few more settlers had gradually moved to Kekopey during the past three years, setting up their houses and planting their fields in the middle of the baboons' home range. To begin with there had been no conflicts, and the baboons had shown only slight curiosity about the new developments.

Earlier on, there had been unexpected trouble in another area. A few of the new transfer males from Cripple Troop, accustomed to life around the army base that bordered Kekopey to the east, had carried their bad habits into Pumphouse with them. Now that they had no army barracks to raid, Chumley, Duncan and Higgins had their eyes—

and very soon their hands—on the small tin huts in which the cattle herders lived. The flimsy walls, insecure door and the gap between the hut and the ground provided ample opportunities for the "bad boys," as we had dubbed the chief male raiders, to gain entry. Generally there was nothing much inside the huts, but the occasional bag of *posho* (milled maize flour) made it worth a try.

It was these raids by the "bad boys" that brought about my first difficult decision and a major change in my orientation. Almost imperceptibly over the last three years I had permitted myself and the other researchers greater leeway in intervening on behalf of the animals. The doomsday mentality that had been born when Kekopey was sold and that continued to haunt the project meant that an occasional monkey was rescued from a half-filled water tank where it might otherwise have drowned. Social interaction between baboons and observers was still forbidden, but it had become more difficult to know where to draw the line about helping an animal in distress. We intervened only rarely, and a "saved" animal became officially dead as far as our project records were concerned, as well as in any of the analyses where such an incident could make a difference in the conclusions. But all this was nothing compared to what I was thinking of doing with the "bad boys." If I left the situation alone to run its course, I also risked their bad habits being picked up by the innocent members of Pumphouse. The alternative was to sacrifice the few for the many.

This possibility would have been unthinkable in the early days of my research, but now the baboons' future seemed to rely on maintaining good relations with the people of and around Kekopey. As if to reassure me that my difficult decision was the correct one, the day before we set up the traps to capture the three culprits, Duncan disappeared. The only news we could gather indicated that he had been killed in the process of raiding.

Chumley and Higgins were easily captured and released as far away from temptation as was possible. I should have been pleased, and of course I was happy that the troop now appeared to be safe. But the relief was mingled with a great sadness. The baboons had trusted me and I had betrayed them. The expressive eyes that met mine as Chumley sat alone in his cage for a whole day waiting until we had captured Higgins seemed to say the same thing. I knew this was only temporary captivity; I knew the males would soon be free and facing a better life than if I hadn't acted, but *they* didn't know it. I stood accused. Being their protector sometimes meant acting like their enemy. I little knew how often I would have to be both enemy and protector during the next four years.

––––––––––

The first news of the extensive crop raiding arrived in September 1979, when I was in California. Five Pumphouse baboons were dead as a direct result of their plundering; by the end of a few weeks, the death toll had reached eleven and I was in utter despair. Each death was a serious personal loss: Paul, Big Sam, McQueen, Naomi, Kate, Frieda and many others. Being so far away made the situation worse. I was powerless to do anything but open the next letter and read the new list of casualties.

The waking reality was bad, but the nights were worse. I had always dreamed about the baboons, probing and seeking answers to my most important questions, sometimes even getting a hint of an answer. But now I was dreaming about delegations of baboons trying to find me, pleading with me to save them. They were afraid, helpless, and they were depending on me.

Commitments I couldn't break kept me away from the baboons for nine months. Fortunately the major crop season was over, and the resident graduate students and I had agreed upon what seemed like a workable arrangement. People were hired to chase the baboons away from the fields and keep them from entering the farming area, even when there were no crops. This was easier than it sounded because the settlement was concentrated on the plateau above the baboons' sleeping cliffs, which meant they were forced to the unoccupied plateaus below.

Why had the baboons suddenly begun to raid the farms after peaceful coexistence for several years? It was only in retrospect that the reasons became apparent. The long drought of the mid-1970s had broken in 1977, at just about the time the first settlers arrived and began to farm. The rains brought an abundance of the baboons' natural food as well as plentiful crops. The monkeys were so wary of people, except for those they trusted—those peculiar baboon watchers—that the real or imagined risks of foraging among the crops held them in check. They did not need the food, and getting it involved too many problems. But 1979 was a bad year. Earlier, when times had been difficult, the troop had managed to survive by ranging over a wide area for their food. Their success depended on having free access to an area covering more than twenty square miles.

Now there were obstacles to that freedom. The fences the farmers had erected around their *shambas* were no deterrent, but the dogs and people who lived around the five-acre encampment were. These farms were the baboons' traditional home range. How could monkeys know that the crops did not belong to them? Weren't they just bigger and better versions of hibiscus flowers or underground corms?

On the other hand, there could have been special reasons that impelled the baboons to try something new, and it was this that I wanted to explore when I finally returned in the summer of 1980 to take stock of the situation. Bob Harding, who had made a recent visit, pronounced the situation hopeless and withdrew from the project.

Even though we'd gotten rid of Higgins, Chumley and Duncan, we seemed to have acquired a new set of delinquents: crop raiders. Cripple Troop once again supplied Pumphouse with a number of fearless, intelligent males who had developed a taste for the easy life when they discovered the garbage pits of a nearby army camp, and now took every shortcut they could find when it came to making a living. Most often, only a few young adults and adolescent males (from about six to thirteen years old) formed raiding parties; older males and other group members were much more conservative. At full force, up to seven males would approach an unguarded field; if the maize was of sufficient height, they would sneak in and wreak havoc before anyone could sound the alarm.

The little gang would leave together and return hours later, looking like clowns in whiteface, with maize milk—the juice from inside the tender kernels—sticking to their faces. Once back with the troop, they didn't do much except rest and sleep. Sometimes they assumed an unusual resting posture, lying flat on their bellies, cheeks to the ground and bottoms up in the air, much like human babies sleeping on their stomachs. Small wonder; they'd eaten thirteen or more maize cobs apiece.

Soon these raiding males attracted a few more Pumphouse members, the first being young adolescent female friends of the raiders who broke with the general female pattern—conservative wariness—and joined their friends. Among the adults, only a few consort females could be persuaded to leave the troop and head for the *shambas,* but when they did they often took the lead, as if whatever had previously held them in check was now gone and they had to make up for lost time.

The damage to crops was greatest when the maize was unripe. Then a baboon would dexterously tear apart ear after ear to check the condition of the cob, perhaps ripping off one and discarding it after a single bite. Thus the raiders left a path of destruction much in excess of what they consumed. When the maize was ripe, however, they would sit and eat as much as they could, stuffing their cheek pouches and creating incredible bulges. Then they would awkwardly attempt to carry away as many cobs as possible.

The older Pumphouse males and the long-term residents were surprisingly uninterested in raiding. Fortunately, the troop wasn't desperate. Being part of a special zone in the middle of the single-rainfall regions of northern and southern Africa, Kenya has two rainfall seasons. The first rains might fail or be insufficient, but luckily there is only a short wait until the next rainy season. This unpredictability also meant that *any* month could be the rainiest one of the year on Kekopey, although rain peaked usually in November and again during April to June.

Even a little rain makes the world a better place for baboons. The new grass and herbs—still too short for zebra, impala, gazelle and other ungulates to consume—the brief flowering of the acacias and other trees, provide a rich variety of food, even if it is only temporary. Although the baboons had first begun to raid during hard times, these periods had not lasted long. The keenest of the raiders, however, seemed to be motivated by more than dire necessity. Habit? Tradition? Did they know something that the rest of Pumphouse had yet to discover?

Whatever the reasons, the crop raiding had to be controlled. I started out with a bag of tricks, some as old as the problem itself. Discovering why such tricks did *not* work might help in my search for something that did. Chasing the baboons was one method, using dogs was another. I planned to try explosives placed around the farms toward which the baboons generally headed. Playing baboon alarm calls and trying different chemical repellents were among other new techniques. I was desperate, even willing to try the local white-paint theory: paint a baboon white and return it to its group and the entire troop would run off—or so the farmers said.

When Bob Campbell had filmed Pumphouse for the Survival Anglia TV special in 1978, he had made some wonderful tape recordings that included baboon alarm calls. Could these sounds, played to the troop at exactly the right moment, be enough to make them think twice about raiding?

The results were ambiguous. Some Pumphouse members were upset, but others were unmoved. The taped alarms seemed more effective when used in a thicket instead of out in the open. Possible reasons for these unexciting results could have been that the tapes were both too short and too old. I myself couldn't recognize more than a few voices, but the baboons had perfect recognition. The taped alarms could have been given by males either no longer with the troop or dead, or by members of specific families, in which cases only members of those families might react. Originally I'd had the idea of using the recordings because of the troop's reaction at the time Bob was making them; I'd asked him to play one recording back immediately, but had to ask him to stop because the troop became so agitated when they heard it.

It was possible that alarm playbacks might work in crop-raiding situations, but to discover how and when would take a great deal of research. I tried a different tack: chemicals. I had planned to collect lion and leopard dung, which is supposed to deter ungulates, and place it around the fields. While I was waiting for the dung to be collected, I tried several more readily available chemical deterrents.

H.A.T.E. C4 is a chemical that had worked well with deer, elk and moose in North America and Europe. It even seemed to have some effect on elephants. It certainly was worth a try, but the cost for the area we wished to protect was phenomenal, and I thought it wise to do my own dry run. A colleague had also suggested "essence of chili pepper," which has been successful against deer when applied to fruit trees. Perhaps it would work with baboons as well: they would handle the peppered maize and get the irritating chili into their eyes, mouths and other sensitive areas.

I descended on the Institute of Primate Research in Nairobi armed with small samples of both deterrents. The Institute had kindly agreed to let me use some of their caged baboons as guinea pigs. H.A.T.E. C4 had no effect whatsoever. These well-fed captive baboons ate the doctored maize with zeal, but the chili-pepper-treated food stopped them in their tracks. So I returned to Kekopey armed with gloves, face masks and enough "essence of chili pepper" to keep Mexico solvent for years. The research assistants and I paired up *shambas* according to the frequency with which baboons were raiding, planning to test both strong and weak concentrations.

Our answer came all too soon. The troop arrived while we were spraying the first farm with the strongest chili concentrate. They gobbled down the treated maize as if nothing had been done to it. We finally created a solution a hundred times stronger than both the company that produced it and our initial trials warranted, but by doing this we ran out of the stuff after barely covering three farms—just enough for a trial run.

As luck would have it, these experiments also ended ambiguously. One farm harvested the remainder of its maize the next day because its owners were anxious about the baboons hovering nearby. Another farm was not raided for two weeks, during which it rained heavily and left us uncertain as to whether the sticking agent for the chili really worked. Both treated and untreated plants were attacked. The report from the final *shamba* was confusing, making it hard to determine which plants the baboons had eaten. So much for "new" techniques. I decided to try some of the traditional methods used to control raiding.

The baboon controllers or "chasers" were still on duty, and some of the farmers were cooperating in chasing the baboons away from their *shambas*. Chasing did work, but only when it was constant or combined with great vigilance, with "chasers" guarding the fields or the entry to the farming area. Again and again I saw the same scene: a gang of males, sometimes even the whole troop, would respond beautifully to being chased. They always left, but they also always waited at a safe distance for another chance. If the humans in charge gave them the chance, they would return to the fields like lightning.

The monkeys seemed to have incredible patience. It was almost as if they had nothing else to do—which meant that the humans had to be equally persevering. Dogs helped immensely, especially if the human guardians were women or children. Pumphouse reacted to men with great fear, were more relaxed around women and almost teased the children. But a dog or two, left free to nip at the outskirts of a baboon phalanx, or to tree a female or youngster, made the troop take greater notice—until the day they discovered three dogs tied up within a compound. Somehow, in the inevitable confrontation, the baboons killed the dogs. After that, dogs ceased to be the powerful force they had been previously.

I was getting desperate. The thought that we might have to do some selective shooting appalled me. It was one thing to banish the "bad boys" to another neighborhood for their own good and that of the troop, but could I myself be the vehicle of their destruction?

Many Pumphouse members had died as a result of raiding, either directly or indirectly. The original death count had been nineteen animals in the first year of conflict. Nineteen eighty was better, because by then we had the farmers' cooperation: only six animals were lost. Four were infants who disappeared for unknown reasons; two were males who might have either been killed or transferred. A rising mortality rate was reported in 1981, probably as a result of the use of dogs and other methods of attack by farmers. The problem was that from the beginning the baboons appeared not to learn that the deaths were directly related to the raids. Why? They learned so much from one another, how could they not sense what was going on?

On the other hand, there was little to notice. A male often leaves the troop and never returns. It happened before there were crops on Kekopey. How could the troop know the cause of disappearance? Did they assume that a male killed while raiding had transferred to a distant troop? Too frequently, these deaths went unobserved. Even when a dog caught and killed a young baboon, it could occur in the midst of such pandemonium and commotion that it might pass unnoticed. If I carried out my deadly plan, any baboon would have to be shot while the rest of the troop was looking. It was the only way the sacrifice could be made worthwhile. I hated the whole idea.

——————

There was a great deal more to consider. Pumphouse had just split into two troops. It had taken three months, and at the early stages what was happening was not obvious, except that this division of a primate group did not follow the pattern I had read about in studies of baboon or other primate species.

At first there had been a typical raiding pattern. The young adult and adolescent male raiders and their adolescent female friends left the main troop to pursue their primary interest—raiding. Soon they slept as a separate group, although always near Pumphouse. Then older sisters, nieces and nephews, even a few of the mothers of some of the delinquents, were drawn into the splinter section.

In most group divisions among primates, whole families choose one side or another. With a few males, these make up the daughter troop. But this time it was different. Families were shattered: some children went with the departing group, the rest with the main section, or mothers in one and some of their children in the other. A few youngsters and adult females were obviously confused. Mother might be in the little new troop, offering all her support—but *that* troop contained no youngsters, and play was possible only in the other, larger troop, where there were plenty of young animals. The young baboons would go back and forth trying to decide which was the most important, play or nurturing.

Beth, the lowest-ranking matriarch, was having problems too. She must have seen some advantages in being with the new troop: she would still be on the bottom rung, but the bottom there might be better than the bottom in Pumphouse. Her adolescent and juvenile sons liked Pumphouse and refused to budge. Only Beth's adolescent daughter, Beatrice, joined her in her indecision.

Soon the two separate troops slept farther and farther apart. They would meet during the day, but relations were definitely cooling. It became more and more difficult for the indecisive members to slip from one group to another. Open hostility broke out. The two troops began to mob each other. This was usually initiated by the small, new troop, but it was difficult to tell which group was dominant. I expected Pumphouse, the larger of the two, to hold sway, but at this point nothing was definitive.

What was certain was the cause of the split: there had been a disagreement about where to go and what to eat. There were three opinions: males in the new troop wanted to do nothing but raid. Some females seemed attracted to these males and were somehow persuaded by them to join in. The remaining members of Pumphouse were the reluctant raiders, objecting strongly to unnecessary raiding when natural foods were available. The name of the new troop became obvious: Wabaya, meaning "the bad guys."

Why raid when you don't have to was one key question. The other important issue was how to explain the reticence of some of the troop and not the others. My crop-raiding research had now expanded to involve three troops, Pumphouse, the newly formed Wabaya troop and Eburru Cliffs, the adjacent troop Sherlock, Ian, Handel and many others had joined. These three troops had all reacted differently to the changes on Kekopey. Wabaya were constant raiders; Pumphouse raided when necessary. Eburru Cliffs went looking elsewhere for the food they previously got from the area that was now *shambas.* In fact, Eburru Cliffs often trekked nearly twenty miles, crisscrossing an adjacent large ranch as they searched for food. While they had found many good locations, they seemed reluctant to give up their old Kekopey haunts. Often they raced back to favorite Kekopey sleeping sites just in time to oust other encroaching troops, only to get up early and make the long return journey to that day's source of food.

Not only did some troops lean toward raiding while others pressed for alternative solutions; there could be raiders and nonraiders within a single troop. Even Wabaya had its nonraiders (by Wabaya standards); Pumphouse had its raiders and Eburru Cliffs included a few males who took advantage of opportunities to raid.

We began to piece the picture together. *Raiders were different*. They spent much less time feeding and more time resting; as far as females were concerned, there was much more time for socializing. The maize was not that much better nutritionally than the baboons' natural diet, but it was a much larger package. Succulent little corms probably contained more protein than a maize kernel of the same size—we were analyzing the two to find out—but the baboons who fed predominantly on corms while their compatriots were raiding the maize fields had to work considerably harder for their nourishment. A corm is the size of a premature baby pea, and extensive preparation is necessary before it is freed from surrounding earth and roots. A large bite from a maize cob might equal more than a hundred corms.

Eating crops saved time. Crop raiders could sit around all day waiting for the right opportunity, and as long as that opportunity presented itself, they were home free. When fields were properly guarded and baboons were excluded for long enough each day, they left. If they were to get enough to eat that day, food had to be found somewhere else. Simply chasing the baboons wasn't going to be enough; only chasing plus guarding could save the crops. Adding some life-threatening factor to chasing—dogs, men or guns—upped the stakes enough to change baboon minds. Then the overriding consideration became risk.

The "activity budgets"—how an individual spends its day—of raiders resembled one another even across troops. Pumphouse male raiders had a budget almost identical to Wabaya males but significantly different from the nonraiding males of their own troop. This pattern held true whatever the troop, season or time of day.

This helped to explain why the baboons did not appear to be interested in the beans some of the farmers were now growing. Beans were small and more time-consuming to eat, too similar to their natural diet. Why take the risks of raiding when the profits didn't warrant it? If this was in fact the way the baboons were thinking, they might become less keen on raiding if less appealing crops were planted, as long as there still were other, natural foods to eat.

For a time it seemed as if there might be a natural solution to the crop-raiding problem, so long as the number of farms remained stable. The rainfall cycles could keep most of the monkeys from becoming habitual raiders. In 1979, for instance, Pumphouse had eaten corms instead of crops. When years of good rainfall alternate with dry years, corms and crops abounded simultaneously. With a reasonable amount of crop protection by people and dogs, crop damage could be kept at a minimum during those times when it would be most important economically: during the good harvest of the rainy years. In the dry years the Kekopey farmers would have nothing to harvest, but the baboons could still damage the struggling plants, so that it would be a matter of guarding to keep baboons in line and educating to keep farmers' expectations reasonable. The farmers already knew when the monkeys' actions were critical or when they simply added to a host of other natural factors that resulted in crop failure.

The baboons were telling us that raiding wasn't inevitable. Moreover, most of them were reluctant to take the risks associated with this new way of life. It might be possible for people and baboons to coexist on Kekopey as long as the monkeys still had somewhere else to go besides farmers' fields and something else to eat besides crops, and if the farmers cooperated in safeguarding their own fields. Time and again, as soon as the need for raiding disappeared—that is, as soon as there was even a small amount of good natural food—Pumphouse reverted to its old nonraiding way of life.

––––––––––

The baboons were cooperating, but nature wasn't. The rains failed for season after season during the years 1980 to 1982, and the baboons became desperate. Even Eburru Cliffs, still staying with their almost migratory pattern, occasionally began to raid as a troop. It was time to try a new plan.

Taste-aversion conditioning is a controversial technique that has proved extremely effective with wolves, coyotes and their prey, sheep, and even with some predatory birds and their prey. Would such a conditioning technique work for primates? Primates and carnivores are very different in their feeding habits. Carnivores gulp down their food, are unselective about what it looks and smells like and are not very finicky eaters. Primates are gourmets by comparison. They select food according to complex sight, smell and taste cues. They can have a highly varied diet, like baboons, who eat many different foods, or be extremely selective, like gorillas, who eat a great deal of only a few favored foods. Would it be possible to fool a primate into eating a "poisoned" food?

Most of us have experienced taste-aversion conditioning—food poisoning can cause it. It is a built-in, evolutionary safeguard that protects us from eating bad food. We would try the same principle on the baboons. If treated coyotes could stop killing sheep, perhaps baboons could be made to stop eating maize.

The initial idea had come from a graduate student during the "bad-boys" era. She had tried several times to administer maize meal laced with the emetic lithium chloride to Duncan, Chumley and Higgins. The idea was explored a bit further at the Institute of Primate Research, but Debra Forthman-Quick and her husband, Bronco, made the real effort with baboons. Deb had trained at the psychology lab at UCLA which had pioneered the technique, and she was already quite expert. She came to Gilgil in 1981 to tackle the project for her doctorate, knowing in advance that it might be impossible. I was confident that if it could be done, these two would do it.

They certainly did the best that was humanly possible, given the circumstances. The only trouble was that after testing a variety of drugs with emetic properties on the captive baboons, lithium chloride had been the only one that would make an animal sick but not kill it, and wouldn't do any damage if the animal didn't actually vomit. But lithium chloride is a salt; and the dose needed to be effective was so high that the animals could detect its presence and would avoid eating the treated food.

To try out the feasibility of the technique, Deb and Bronco gave it a field test. Half the troop was trapped in its home range, using maize as bait. They were injected with a sedative and then with the lithium chloride, and finally brought back to consciousness with an antidote to the sedative. The animals were released and then watched closely. This procedure was not what either Deb or I had had in mind in the beginning. I was searching for techniques that would be simple and easy to use by the relatively uneducated farmers. The trapping technique was expensive and sophisticated, but it would allow us to find out one important thing: Could baboons develop taste aversions to maize? If they did, we would have to search for other chemicals or other ways to make the animals sick.

The first stage was completed successfully. Since the bait that made the baboons ill was maize, it was maize they associated with illness. This was what was needed. The next step was to find out whether an aversion had been created. As luck would have it, 1982 was a drought year—so bad that there were no crops to tempt the baboons, and none with which to test their reactions. Deb was forced to test her subjects with a variety of foods that she had to present to them on the sly. The majority of those animals who had undergone the lithium chloride treatment showed some evidence of an aversion to maize, and more than a quarter even had a strong aversion. Those who had not been treated devoured potatoes, carrots and maize with equal relish.

The results, while encouraging, could have been better. The reasons were obvious. First, the experiment had been under a handicap to start with. Maize has very little flavor, which makes developing an aversion to it more difficult. Furthermore, the animals had eaten and enjoyed maize many times before, and a single bad experience might not overpower many good ones. The more often an individual becomes sick, and the stronger the reaction is, the more intense the aversion becomes. Because these were very valuable baboons and needed to be protected, a weak solution of lithium chloride had been used, and it was used only once.

Deb's work did suggest that aversions could be created in a primate, and could last at least five months, the longest period she had available to test the baboons' reactions. But there were limitations. The animals didn't learn to generalize from their unpleasant experience—the baboon who wouldn't eat bananas at IPR because of unpleasant taste-aversion reactions still gobbled down banana *skins*. The animals did seem able to generalize maize kernels from maize on the cob. This kind of generalization would be important. For example, what if an aversion to young maize cobs was created and the baboons simply waited for the cobs to age before they ate them?

Deb had other observations that were more interesting and also more devastating. I knew from my own observations of wild baboon feeding habits that monkeys respond to social pressure. Deb tried feeding maize to a captive male with a known maize aversion in full view of nineteen other hungry baboons, many of them his seniors. These onlookers wanted the maize, and threatened loudly and vigorously. Oblivious to what his stomach was telling him, the treated monkey quickly ate half of the maize cob. When commercial monkey chow was offered and the onlookers lost interest, the test monkey switched foods. Later, however, he returned to the maize, examining it slowly and more carefully, sniffing and biting off kernels; he then threw them all away, returning to feed on the chow. Clearly there is not that much difference between humans and baboons. Even if you don't really want what you're offered, it must be worth having if someone else wants it badly.

Of all the techniques we tried, taste-aversion conditioning seemed to hold the greatest promise. Granted we would need to find a new chemical—preferably one that was tasteless to a primate palate, that would produce a strong reaction without being lethal and that could be easily administered more than once. A tall order, but not impossible.

Meanwhile crop raiding was increasing. When times were rough, more baboons raided. They had no alternative—there were few other foods and their range was restricted by the new human occupants. But even when alternatives did exist, Wabaya raided. What did they gain? They saved themselves some time, but what did they do with it? They rested, and a few individuals, mostly females, socialized more than usual. The advantages seemed slight.

But as time passed, I began to notice more advantages. It might not be a coincidence that adolescent and young males were the initiators and prime raiders. These were the animals with the greatest energy needs because of their growth and size. They were also the ones who might sustain the greatest conflicts between feeding and reproductive activities. But the raiders did not seem to combine the two; they opted out of normal male competition for females to devote all their time to raiding, and as a result seemed to grow faster than males eating natural foods. Then—if the example of a few males was representative—they'd stop raiding and return to the group. Because they had grown more quickly, they could enter the reproductive arena at an earlier age, gaining an advantage over the more conservative, non-raiding males. Perhaps a young male could manipulate his overall reproductive success by shifting his foraging strategies. Could this be an evolutionary response to improved resources? In this case, the resource was manmade (crops), but in other cases it could be part of natural cycles.

Wabaya females also seemed to be gaining an advantage. They were certainly in better condition, with thicker coats of a reddish tint resembling those of Cripple Troop, rather than the usual grayish-brown color. Originally I had thought the Cripple Troop color was of genetic origin; now I was less sure. The most telling difference was in the rate at which they produced babies. While Pumphouse females had infants about every eighteen to twenty-four months, and Eburru Cliffs females at even longer intervals, Wabaya females now produced babies every eleven or twelve months, a clear indication that they were in better condition. Perhaps in the end they didn't fare so well, since injuries, deaths and disappearances related to crop raiding took a major toll of the babies and even some of the mothers. Raiding had created a new and dangerous environment for baboons.

The raiding persisted, and the farmers became angrier and angrier. My academic colleagues were of no help: since the conference, I was no longer on speaking terms with some of them. I didn't know where to turn, either for practical help with the crop raiding issues or for moral and intellectual support for my own plummeting spirits.

It was at this crucial point that I renewed my acquaintance with someone who was to play an important role in my private and professional life. I had first met David (Jonah) Western in 1973. He was a strikingly attractive man whose sandy hair became a light blond when bleached by the sun; his skin was golden brown, and his rugged appearance contradicted an almost shy demeanor and a razor-sharp intellect. Jonah, his nickname from school days, was already a legend by the time I first arrived in Kenya. In 1967, he had come to Amboseli to study the causes of the drying up of an important ecosystem, a study that immediately placed him in the midst of controversy and conflict. Were the Maasai, the local pastoralists, degrading the area through overgrazing by their domestic stock? Were the elephants to blame? What action, if any, should be taken? Even in the mid-sixties, these were not unemotional issues.

Jonah was a keen naturalist and an exceptional scientist, and he unraveled the mystery in a way no one had expected: the blame lay with a natural cycle. When the water table rose, as it did about every thirty-five years, the soil salinity increased and killed the existing species, including the yellow-barked acacia trees. The umbrella acacias also suffered, as did the grasses. In their place came salt-tolerant vegetation, and the whole landscape was transformed, plants and animals alike.

But Jonah was interested in more than purely academic matters. His childhood in Tanzania had instilled in him a love of animals and wilderness, and he was dedicated to conserving both. His vision of conservation was unusual. It included the foundation of the coexistence of people and animals—how else, he reasoned, could there be a long-term future? Research led him into planning, which in turn led to an even greater involvement in government policies. At the age of twenty-five, in 1974, he played a critical role in the establishment of Amboseli as a national park. By the time he was twenty-seven, he had secured a multimillion-dollar loan from the World Bank for the development of Kenya parks. By thirty-four, with the aid of the Canadian government, he had set up the Wildlife Planning Unit for the Kenya government.

I had known Jonah casually for many years. Monkey watchers were not really considered part of the Kenya wildlife scene, but our paths crossed more and more frequently. Our conversations usually concentrated on research, and though in our first meetings I mumbled with nervousness, Jonah's quiet demeanor gave me courage. We explored the parallels and the differences between our work and our worlds until Jonah's approval convinced me that my ideas about baboons were reasonable, and I shouldn't be timid about saying what I thought. His calm, determined support gradually worked its magic; he was my opposite, as socially shy as I was outgoing, as intellectually bold as I was reserved. In a practical sense, his conservation insights gave me the added push I needed to tackle the farmers in an effort to find a solution to the crop raiding and to save the animals from destruction.

The raiding situation was at an all-time low. Morale was bad among both the researchers and the farmers on Kekopey. The year's raiding had not yet begun, but there was leftover ill feeling from the previous years' troubles. The air was charged with tension and only slightly veiled hostility toward both baboons and baboon watchers. Trying to decide who was responsible for the baboons seemed to be the greatest problem.

In Kenya, wildlife belonged to the government—that is, to the people of the country as a whole. Even on private land, the wild animals were only held in trust by individual landowners. The government decided the fate of the game, in both senses of the word. National parks excluded people in favor of wildlife, but even there the rights of humans were considered. When Amboseli became a national park, the displaced Maasai were compensated in a variety of ways, both initially and for the subsequent depredations caused by park animals outside the park boundaries.

Legally, the baboons belonged to the government; the government represented the farmers, so the baboons belonged to the farmers. Even without the last link, this was a difficult concept to get across. At first

I thought the settlers were being obstinate, but slowly, with Jonah's help, it dawned on me that what they saw and what we were telling them didn't fit together. They saw us walking with the baboons, usually behind or in front of them; when the troop headed for the *shambas*, we were always either rushing ahead to warn people or desperately trying to keep up with the monkeys. It all looked familiar to the farmers; didn't goatherds, shepherds and cowherds follow behind *their* animals? To the farmers it looked as if we were herding the baboons; worse still, it looked as if sometimes we actually *led* them to the crops.

What I was now actually confronting were the problems of cross-cultural communication. It made me realize that cross-species understanding could often be more easily accomplished than overcoming cultural barriers. We had to convince these people that the baboons were wild and that we couldn't control them. Since the baboons didn't belong to us, protecting the crops from them was not our responsibility; it was that of the people who actually owned the land and the crops.

But no matter how many times I discussed this with the farmers, the response was always the same: total incomprehension followed by angry accusations. What was I going to do about the crops? The baboons? The whole problem?

By now I hated the farmers. My first encounters with them had been when I was out watching the baboons, before I had really faced the issues. They came to gawk, to disrupt, to air their grievances. My reaction frightened me. My basically unaggressive self was transformed, and I wanted to kill—the farmers. It was only in retrospect that I could see hatred and ignorance working together. I didn't know the farmers; they were the faceless enemy. The world was divided into good and evil: the baboons were good, the farmers were evil. I was on the side of good, the side of the animals, and the farmers were on the other, bad side, responsible for the destruction of the baboons, both immediately and ultimately.

Jonah listened to me patiently, then slowly planted his ideas. People were part of the problem, so they must be part of the solution. They had to protect their own crops by being vigilant and chasing the baboons away. The troop was still frightened of people, and even a small effort on the farmers' part would send the baboons rushing for safety, away from the disputed crops. But the farmers weren't playing their part. People were rushing *into* their houses, not *out* of them, and locking the doors until the baboons left. These settlers all came from places where there were no wild animals, and they were more frightened of the baboons than the baboons were of them.

I was angry and discouraged. I had heard many rumors via the grapevine—that the farms were bought on speculation by rich Nairobi people who were absentee landlords, using relatives or hired help to tend the plots; that the people were dishonest and would cheat as much as they could. It was only my concern for the future of the baboons that prevented me from simply walking away. I had chosen not to be a cultural anthropologist for the very reason that I didn't want to work with people and alien cultures; my years with the monkeys had reinforced this. Yet here I was.

As a first step, and again with Jonah's help, I organized a *baraza* or public meeting, and enlisted the aid of wildlife and local officials—anyone who could be useful. It was an attempt to open up channels of communication, to try to discuss the problem and its possible solutions. The meeting was a success, partly because I was able to involve the Wildlife Clubs of Kenya, an important and effective conservation organization with an excellent and sophisticated staff. They assisted me by providing a sympathetic interpreter and later sent a mobile film unit to begin a program of wildlife education. Trips to nearby Lake Nakuru National Park were also arranged by the Wildlife Clubs; it was the first time these Kenyans had even been in one of their own national parks. Speeches were made in Swahili, the national language, and in Kikuyu, the language spoken by the farmers. But our ace in the hole was *harambee*. *Harambee* is an important Kenyan concept, was the watchword of Jomo Kenyatta, Kenya's first president and the slogan of Kenya's independence. Loosely translated, it means "Let's all pull together"—to help a solve a community problem, to build a school. When a school is needed, parents and friends contribute as much as they can; this gesture is a prerequisite of any government aid or support. When someone falls on hard times, he invites his neighbors to share a meal he can ill afford. In exchange, they will be expected to contribute money to help him out. In this context, *harambee* meant that we, both farmers and baboon watchers, had to work together to solve our problem. The *harambee* meeting itself eased tensions, and although the problem remained, spirits lightened. It seemed as if the farmers were willing to join in the efforts to find a solution.

Then the raiding began again. It happened quietly one day when part of the troop slipped past the chasers and found some newly ripened maize. At first only a small subgroup were raiding, but soon the entire troop followed their lead. Two weeks into the harvesting season, Kekopey was a war zone. There was one difference from the last raiding season: this time the farmers really were doing their best to solve the problem. They chased baboons and their dogs chased baboons, but often the monkeys were too smart, too persistent and too many. What could we do?

# 14. Humans

Jonah's words echoed in my mind: "People are part of the problem; they must be part of the solution."

I set out again, grimly determined to find out what the human part of this deadly equation was up to. The least I could do was to assess the amount of damage the baboons had done to the *shambas*. The *baraza* had cut the ice, but I still perceived the farmers as the enemy, the serpent in my baboon paradise. They were stupid, dishonest, avaricious human beings, and I secretly hoped they would miraculously disappear from the face of the earth—or at least from Kekopey. But the twelve hours that followed changed my mind and my life.

It was September 1980, and as I went from farm to farm, from person to person, I discovered that the destruction was extensive and that the truth about the farmers was far from what I had imagined. For many of them, this was their first piece of land, and it was obvious that owning land meant a great deal to them. For the most part, and despite some farms that were owned by absentee landlords whose relatives worked in exchange for shelter and food, they all were raising crops for their own consumption.

I was deeply touched by the people I met, by their courage, hardships and attitudes. And I was surprised at their honesty. I had done my homework about crop yields under similar conditions, and with only two exceptions no one exaggerated their losses, even though it would have been to their benefit to do so.

The second day was much like the first, with one important differences. The generosity of the people overwhelmed me. Obviously, the farmers were pleased that I took the time to come and see the problem for myself. Whenever I was ready to leave, I was laden with gifts of food. I protested, exclaiming that I couldn't take the little that remained after the baboons had blitzed their farm. But they explained that baboons were like thieves—I should take what was left before it too was stolen. Today I wasn't being held personally responsible for the actions of the baboons.

As I struggled home, laden with my clipboard, several cobs of maize, a pumpkin, two sweet potatoes and a watermelon, I had a great deal to think about. Since I knew the land was too arid and rocky for much farming, I had assumed that the new owners had been fooled into buying it and must be dissatisfied with both its quality and its price. Surprisingly, this was not so. The farmers felt they had been treated fairly, and if they could raise the money, they would buy more land. At one time I had entertained the thought of a mass relocation of settlers as one solution to the current difficulties, but apparently no one wanted to move.

I gradually came to know and even like some of the farmers. One I shall always remember is Rosemary, whom I had mistakenly called Rebecca on many of my visits. Kikuyu names are difficult: each person has three names that are used at different times by different members of the family. Rosemary's farm was near a favorite sleeping site, an easy target for the baboons. Shortly before one of my visits, the monkeys had finished off what remained of her maize crop. When she told me what had happened, she showed none of the anger of the farmers who had come to me at the Red House with their complaints.

Rosemary and her neighbors lived humbly, for the most part in houses built of rough lumber with corrugated tin roofs. Other houses were even more traditional, of hard mud coating a frame of twigs and branches, topped with a thatch of grass. All had windows and doors cut into the walls, without glass but with wood shutters. Several enclosed a dirt courtyard. Floors in all the houses were of hard-packed dirt kept spotlessly clean despite the goats, chickens, sheep and dogs that wandered through.

Everyone took great pride in his or her home; they had built them themselves. Although each family seemed poor by Western standards, for a Kenyan to own a piece of land is considered the greatest asset. These subsistence farmers had only recently entered a market economy where cash was required. No matter how bare a house might be, the family living there owned a transistor radio, which always wore a hand-embroidered dust cover. Countless snapshots of family members adorned the walls.

Every house held its own story; most families struggled hard to cope with Kenya's new realities. After only a few generations of medical care, health had so improved that numerous families now contained so many members that the land was overwhelmed. As a result, the younger generation had to look to the marginal regions such as Kekopey, which had not yet been developed agriculturally.

All the farmers and their families were fearful: Kekopey was very different from their traditional home. It was so dry, the rainfall so unpredictable, that no one could guess how the crops would fare. They were on their own, cut off from their extended families and for the first time surrounded by wild animals, creatures that had long ago disappeared from the main agricultural regions of Kenya.

These tin roofs that I had at first found so distressing sheltered many hopes and fears; many ambitions were focused on the success of this new venture. The people were worried: Would harvests provide enough to sell to markets so that they could have enough money for school fees and uniforms, tea and sugar, a little meat? The language of these people was foreign to me, but the feelings were not. I could no longer harbor bitterness toward them.

What was happening in Kekopey reflected what was happening all over Kenya and in many other Third World regions. Better medical care, a lower death rate and a continued high birthrate resulted in a rapidly expanding population. A family's traditional parcel of land, which in times past had been able to support the surviving children, was now divided up into untenable pieces and handed out among an increasingly large number of inheritors. Those in the first, new-style generation might be able to remain on the family plot, but by the second generation the land would be unable to support the growing number of people and a scramble for new land would begin.

The problem was that all the good land had been settled and developed years earlier. What land was available resembled Kekopey—excellent for cattle but marginal or worse for agriculture. Given these options, it was no wonder that land on Kekopey was at a premium and that the farmers were not in the least unhappy with their lot. They had arrived out of necessity. Fortunately no one would starve to death, as would have happened a century earlier. The Kenya of this decade had diversified its economic base, even for subsistence farmers. The extended family formed a hedge against disaster: someone back home would have had a successful harvest, or one of the more educated younger generation would have a job in Nairobi that brought in money for food and other essentials. What people really wanted was security, a piece of land that could be called a home. If you could make a living from that land, so much the better. It was certainly worth a try—and if you couldn't, at least you knew you always had somewhere to go.

Knowledge and contact are powerful tools. My hatred vanished along with my prejudices as I came to know these people. Clear-cut issues became fuzzy. Once I saw both sides, I also saw the tragedy of the situation. These people had a right to the land and its bounty, but so did the baboons. The settlers had nowhere else to go, but neither did the baboons. Each side was acting in accordance with past tradition and reason, even if the two were diametrically opposed.

The new reality existing on Kekopey would not go away, however much I might hope it would. Jonah provided invaluable advice, experience and moral support. Over these years—1979 to 1981—we had become a couple. Discussions on research had led to many other discussions, days spent together were followed by weeks and months shared in California and Kenya while we discovered that despite our different backgrounds we shared a great deal in common. Now we were "in consort."

One area in which Jonah and I did diverge was in conservation and academics. I was fascinated by the intellectual issues associated with the baboons, and years of being with them had produced a deep personal commitment to them. But the Pumphouse Gang was in trouble, and I personally was having to confront many of the basic conservation issues facing wildlife in Third World countries. I hadn't thought much about all this before, and although I believed conservation to be important, my half-formed ideas lacked both knowledge and sophistication. Jonah was the opposite. He had become a pioneer in a new type of conservation. Conservation issues had led to academic work that gave him a better understanding of how places and animals could be safeguarded.

Costs and benefits of wildlife to the people living in wildlife areas had to be considered: there would be no realistic future for the animals if the costs of conservation to the people consistantly outweighed the benefits. One of Jonah's major accomplishments was to show how a management plan beneficial to both animals and their human neighbors could be created and implemented.

It was fortunate for me that Jonah had exactly the experience and sophistication in modern conservation methods that I lacked. And so a new project was born, a new direction and in many ways a new me. For the first time since I became an anthropologist, I became involved with people.

To begin with it was difficult to adjust to this life and this project. The "people business," as I called it, took up more and more time. I saw the baboons less, and when I did they were always either trying to get into trouble or to get out of it. I simply was not emotionally prepared for the change of role and direction. The baboons had always been my inspiration; being with them revitalized me. Now this new work took me away from the baboons. I had had no formal training, it was mentally and physically wearing, and I had only Jonah and my instincts to guide me. The prospect seemed bleak, and often my day would end in tears, not so much of despair as of exhaustion. I was unused to this constant state of crisis. Eventually, however, both the "people business" and the crop-raiding research created new challenges that became intellectually engaging.

———

Baboons are smart creatures, but their response to raiding opportunities showed me a new dimension of their intelligence and for the first time made me wonder if the farmers or I could in fact outsmart them. Getting around the beleaguered chasers was seldom much of a problem. We increased the staff, and had them hurl stones and shout, which made them more effective. Nonetheless, it still took us a while to catch on to some of the monkeys' tricks. Lou, one of the keenest raiders and a transfer from Cripple Troop, original home of the "bad boys"— Duncan, Chumley and Higgins—might look as if he had no interest in raiding as he moved in the approved direction with the rest of the troop. But as soon as he reached the thicket, he would circle back, make a wide arc around the chasers—whose guard was now down, since the baboons were heading out of trouble instead of into it—and before anyone realized it was gorging himself on maize cobs.

Lou wasn't the only sneaky baboon. In general, the young males were the most avid raiders, and didn't mind foraging on their own or in small destructive groups. Sometimes the entire troop followed their lead, sometimes only a few friends and relatives. Even if the farmers were on guard, they often found their task difficult. The baboons were still afraid of them, and would run off when chased, particularly if there were dogs around. But they weren't afraid of women or children. Whenever the baboons left, they ran only a short distance away until the people went inside their huts or were so exhausted by all the chasing that they dropped their guard. The baboons would be in at once; and again and again. One could not help admiring their persistence and their maneuvering. I pitied the farmer who was alone when determined baboons arrived. While he would chase some out on one side, others would sneak in on another.

Neither I nor the research students could chase the baboons ourselves. The best we could do was get to the plot ahead of the monkeys and mobilize the resistance. It was painful to watch the baboons win time after time, but occasionally it was possible to get some distance from the events, so that the scene took on the hilarity of a Laurel and Hardy movie.

The "people" side meant more than just chasing baboons away. Jonah's underlying conservation philosophy held that people had to be motivated to preserve animals out of self-interest.

When the Maasai at Amboseli received a share in park revenues as well as other benefits, they changed from being destroyers of wildlife to becoming unofficial rangers, intercepting trouble and protecting the animals. The system had also worked for other areas in Kenya. What benefits to the farmers would outweigh the costs of the baboons on Kekopey? It was hard to know. Naturally people wanted and deserved compensation for crop damage caused by the monkeys, but this was a government matter and a government policy. All I could do was help locate the forms and speed up the bureaucratic process; I couldn't become directly involved. If I found the money to compensate the farmers, I would be admitting I was responsible for the baboons; how could I then be able to claim that I had *no* responsibility, legal or otherwise? When I finally figured out the amounts involved, it did sound tempting. Less than $2,000 a year would cover all the damage; to be sure, it was money that I didn't have, but compared with other project costs, airfare to Kenya and running a car, it seemed like nothing. Even more appealing was the knowledge that the money would go directly to the farmers, who were hard pressed to absorb the cost of the Kekopey baboons.

Another suggestion of Jonah's opened new avenues in my thinking. Why not compensate the farmers in alternative ways, by helping them with social services or other pressing problems? This might not involve much money, but could be of considerable use; and what was more important, it would be seen as coming from the Gilgil Baboon Project itself, as part of *harambee,* yet would not imply that we were responsible for the baboons' actions. There were distinct possibilities to this idea. Certainly my contacts in Kenya and internationally were better than those of the Kekopey farmers. I could organize a way to give them help with appropriate technology, energy, soil and water conservation, wildlife education, agricultural techniques and crops. Kekopey was much more arid than anywhere else they had lived before. Perhaps maize was not the best crop to grow in an area with such low rainfall— and another crop might be less tempting to the baboons.

A rural development project was born. My aim was to help the farmers help themselves. If, in the process, we could come up with a solution that would produce less conflict between the animals and the people, so much the better. By now, not just the monkeys but *all* the wildlife was attacking the crops, including the eland, warthog, impala, tommies, zebra, smaller antelope and hares. The rural development project had its own evolution, an organic process. It might have been better if a development agency had come in with its own specially designed "grand plan," but while neither I nor the farmers had much experience with rural development, we did have one great advantage: I was familiar with the area and its possibilities, and the farmers knew what they wanted. The solutions we forged together were more appropriate than what might have been imposed from outside.

The new type of baboon research and the addition of the rural development scheme meant more manpower. Again, Jonah offered a good suggestion. Why not hire local Kenyans to do some of the work? I had been reluctant to do this in the past, because I was uncertain how the baboons would react; they'd had many unpleasant experiences with black people. I feared that in the time it took Pumphouse to learn to trust Kenyan workers I might lose months of precious data,

But Jonah's idea made a lot of sense. Bringing in enough students from abroad presented many problems; there were always differences in schedules that threatened to leave the baboons unobserved for some periods. An American or European student invariably had to face a difficult adjustment process, as well as contending with the sheer expense of living in Africa: airfare, the need for some kind of vehicle, stipend. A Kenyan research assistant, if he viewed this as a suitable job at all (and many Kenyans hated living in the bush as much as some Americans do), was committed to it from interest and necessity. He knew the country, the language and how to get around, and would work hard even under adverse conditions because good jobs were hard to find.

Josiah Musau became my first Kenyan research assistant in November 1980. He was an attractive, lean man, and I always thought of him as a tribal elder, even though he was younger than I was. Josiah was a member of the Wakamba tribe, the second largest agricultural tribe in Kenya. He had left school at an early age and been taken on as a boy Friday by an American zoologist who recognized that he had an unusual and active mind. Josiah was an exceptional student; he thrived on the instruction his employer provided, traveling into Nairobi to find library books that would help to answer questions raised in his lessons. Josiah was as educated as if he had received formal schooling, and his American accent made people take notice. I was lucky to have him with me.

The first day I took Josiah out to meet the baboons, I was prepared for the worst, even though he was dressed to look as much like me as possible: backpack, binoculars, long pants and a hat. The only sign that indicated he was not one of the "regulars" were his hands and a small part of his face. My presence was usually enough to reassure the animals that a new person should be accepted, and I hoped their trust would carry them through this challenge.

At first they didn't react at all. It was only when we edged closer that several baboons glanced up, to receive the surprise of their lives. Their expressions showed the shock and distress they felt; they gave alarm calls and ran off in confusion. But *I* was with this person, and *I* didn't appear alarmed. They ran a few paces, then turned, sat down and stared at Josiah. We calmly stood our ground, and by the end of the morning the troop permitted Josiah—when he was with me—to penetrate the edge of the group, although they became extremely nervous when they saw him move on his own.

Josiah had grown up around animals, and was particularly sensitive, moving gracefully and naturally among the baboons. This was in striking contrast to the typical American students, who were always falling into pig holes, shouting and sending clipboard and binoculars flying, and whom I constantly had to admonish to "Watch out for the baboons!" as they barely missed stepping on an infant cowering behind a bush.

---

As project aims grew, the staff expanded. After Josiah came Hudson Oyaro, then Simon Ntobo, then Francis Malele and others. I gradually added more Kenyans—including, of course, the chasers, but also one of the farmers, who acted as a translator and also monitored the planting and harvesting of all the farms. Involving Kenyans in the project was the best thing that could have happened; I wished I had done it sooner.

But despite all this, life at Gilgil had stopped being fun. Crisis followed crisis and difficult decisions seemed to lead to even more difficult ones. A large staff meant that I was more involved with administrative details and spent less time with the baboons. Conflict, too, became a way of life. Had it just been the conflict between the farmers and the baboons, perhaps it would have been manageable, but the baboon problem was at the core of other, more personal troubles.

A remarkable number of new students came to study the Kekopey baboons, not only the Pumphouse Gang and Eburru Cliffs, but also Wabaya and, briefly, Cripple and School troops. Some came as volunteer assistants, some as postdoctoral researchers, but the great majority were graduate students working on their thesis projects. As the baboon situation deteriorated and I assumed a more active role as director, conflict between the students themselves and between me and the students became inevitable. Most of the arguments concerned priorities—who could do what at what time. Later, conflicts turned into competition; the more similar the academic interests were, the more intense the competition.

Each new crisis placed me at odds with the graduate students, whose perspective was very different from mine; most took the short-term view, with their own studies the central and most important piece of work to be accomplished. For them, anything that stood in their way was bad and to be resisted. I had a longer-term perspective. Now that the lives of the baboons were in jeopardy, I couldn't be blasé about even small matters, because to do so could result in devastating repercussions. The golden years of Kekopey were coming to an end; we were no longer a happy family. For the first time I realized that being the person in power, no matter how reluctantly one might have assumed that position, meant making enemies. Of course, there can be hard feelings even when equals disagree, but when two who are not equals don't see eye to eye, there is bound to be bad blood.

Working together, exchanging help and ideas, ceased; stony silences developed. There were peculiar lapses of memory and much obvious manipulation. I was astounded to learn from one student's adviser that he was tremendously excited about his student's work, since it uncovered for the *first* time that male baboons did not have a dominance hierarchy! I was dumbfounded when I received a call from a student who was about to finish her thesis on the baboons. She informed me that she no longer wanted to receive any more of my papers or manuscripts. Until then I had discussed my ideas with her freely. Earlier I had given her project a new and important dimension that she had overlooked. What was she afraid of? The answer was only too apparent when I reviewed her thesis before its final defense and read her subsequent work: there was almost no reference to my papers or to the ideas on friendship, reciprocity and social strategies that I had been developing for over a decade. Now she was free to claim as her own many important concepts that found their origins in my own work. This was not an isolated incident; competition went hand in hand with deceit, conscious and unconscious.

Here was the dark side of baboon research, and sadly it played a large role. Old friends turning into enemies would have been bad enough, but the new directions in which my research was moving brought me under fire from heavier guns. Although I had originally planned to temporarily remove all the male baboons during my initial study, in order to see what the females would do, I never executed these plans. I felt I had learned enough not to intervene. But I had unconsciously also assimilated a certain orientation toward the baboons; I would leave them "natural." What *natural* meant to many of my colleagues and superiors was "untouched by humans." Kekopey had been criticized on this score, since it was a cattle ranch and not a national park or reserve, although there was more wildlife on Kekopey than in many parks, and more people and cattle in many reserves than on this particular cattle ranch.

The point was that in the minds of some important colleagues, pristine, untouched nature did exist. Only there could you study baboons and gain important insights. Once humans intervened, this resource and its potential vanished instantly. It was bad enough that the baboons lived on a cattle ranch; now they were raiding farms and eating "human" food. And I was now proposing to manipulate them further, to subject them all kinds of tests and experiments. If Pumphouse had ever had a chance to be considered normal, natural baboons, if my insights about baboons ever stood any chance of being seriously considered, it vanished with this new turn of events.

---

Are you responsible for what you tame? I had not given this much thought before, but my instincts, actions and now my conscious mind told me that yes, I was. I must ignore the skepticism and the critics if I was to save the baboons. How else could I repay my debt to them? More than that: Until now, I had been a naïve animal watcher, a reluctant conservationist. As I explored other situations, looking around for answers, I realized that ivory-tower academics will have to become more responsible for their animal subjects if they want any animals to study in the future. This problem exists throughout the Third World. As underdeveloped countries try to modernize, and as their human populations expand, the state of local and global economies means trouble for wildlife as well as for wild places.

But could we blame them? Did Western nations have the right to point a finger in accusation at the Third World, when most of those industrialized countries had themselves long ago destroyed their own animals and countryside in the very same rush to develop their resources? Now that *we* were sitting pretty, we wanted to ensure that both ourselves and our children would always be able to enjoy the beauties of nature elsewhere.

Jonah's solution still seemed the best one. Consider the people; work out ways in which they can benefit from animals without destroying all of them. It was a pragmatic and realistic solution, and it was working, even on Kekopey. Although we hadn't solved the crop-raiding problem, we had made considerable progress with the rural development scheme. A new school, modest but adequate, had been built with money from the proceeds of a Survival Anglia film about baboons. Farmers were being sent to courses on soil and energy conservation, tree planting, the development of new agricultural techniques and crops. Wildlife education programs to teach the farmers more about the animals had begun. Government agencies were helping with various aspects of development.

In October 1982, I was presented with a novel idea. Monica Geary, an Australian weaver, had come to Kenya to teach rural people a new means of livelihood through crafts. She was looking for an appropriate place, and our paths crossed through an old friendship she had with one of the Kekopey farmers. Thus began the Woolcraft Project. In three short months, the farmers learned to prepare fleece purchased locally, spin it, color it with natural dyes, create yarn for knitting and weaving and make rugs of remarkable beauty. If the Project worked, the farmers would have an important new source of income, one not dependent on rainfall and one that might even turn them away from agriculture and from their conflict with Kekopey's wildlife.

A comment made by one of the farmers reminded me of how far we had come. He had been the most hostile of those who came to our first *baraza*, the meeting that had originally broken the ice, but now was a friend and ally, a pioneer in trying new ideas. What he said was, "We would rather have the baboons here raiding and the Baboon Project studying them (and helping us) than have no baboons and no project." I couldn't believe what I'd heard, and had to ask him to repeat it.

The benefits we were bringing were indirect, but in the plans for the future were other, more direct ways in which the farmers might stand to benefit from the surrounding wildlife. It was easy to see that people would act out of enlightened self-interest on behalf of animals, but unreasonable to expect them to be altruistic; they were too poor and life was too difficult to ask that they sacrifice any portion of what they had for creatures who were only distantly related to them.

I could have saved myself much time and grief if I had applied my baboon insights to these human problems from the start. If you want to forge a good social relationship, one that works and endures, it must be built on reciprocity. But first you must gain the trust of your partner. Whether you are a new male baboon or a female scientist, you must demonstrate by action and gesture that you are generous and reliable and that you will make a good social ally. It is only then that you can count on your partner's goodwill and can help each other out, not necessarily in identical ways but in complementary ones.

This is why some approaches to conservation cannot and do not work. People are too often ignored, except in devious ways that serve only to exclude or coerce them. It has been argued that what we need are armies to protect our precious dwindling wildlife, but there is not enough money or manpower for such a solution to stand a chance of working. Whenever it comes down to people versus wild animals, the animals will lose if the people are desperate; after all, the animals operate the same way themselves. We all share that incredible urge to survive. The key that Jonah discovered many years ago is to make the two sets of interests coincide: to set up a system of reciprocity between animals and people. It is no small task, as I was at that moment discovering. It required considerable time and ingenuity, and even if it worked, it could not be a permanent solution. Everything changes, and all solutions have to be modified; new approaches have to be developed that can cope with altered reality. Conservation is a never-ending task, but what I was learning made me both more educated and more optimistic about prospects for the future.

The direction my conservation ideas were taking had unexpected rewards. Helping people to help themselves was tremendously satisfying. After so many years of being at a stalemate with my social conscience—why was I so involved with baboons when the world was so full of people needing help?—I now had a chance to help both.

————

The research itself was making progress. My growing understanding of why the baboons were raiding was accompanied by many new insights more basic to our general understanding of such smart animals, and even of human evolution itself. Once again the monkeys had led me down a path I would not have chosen for myself, forcing me to explore more interesting and important issues.

So much had changed or was changing: Kekopey, the baboons, the project and my role in it. I had come a long way—from observer to interpreter to protector. I was too busy to think of what the future would hold, and Peggy's death in November 1982 was a symbol, an indication that the future would be very different from the past.

Throughout the ten years I had studied Peggy and her family, she had been very special to me. When I arrived back in Gilgil that November, after one of my trips to California, Josiah and Hudson came to tell me that Peggy had fractured her leg and was unable to move. She was sitting close to a watering place frequented by the troop. Peggy had had a close call before, while I'd been in California. She'd recovered then, but she was now well into her thirties. I expected her to go someday, but I had never imagined it would happen this way.

By the time I arrived, Peggy had been in the same spot for a week. It had been raining hard, and I worried about her being left in the open. She was extremely weak, and my first thought was to bring food for her and try to set up some kind of temporary shelter. As I was doing this, I discovered that she would let me touch her while she was eating. It was then I noticed that maggots had already infested her injured leg, and realized she could not survive.

At first I did not know if I should try to save her; we had faced difficult decisions with injured animals in the past, and had managed to help some, though each new case required a weighty decision. But this was Peggy, and my emotions dictated my actions. At least I could bring her to the Red House and shelter if she would let me, and better still, take her to the Institute of Primate Research, where they could attempt to treat her. This was an unprecedented move on my part, but since we'd already been blitzing the troop with our attempts to control their raiding, barraging them with noxious stimuli, trying to manipulate them, why should I pretend that life was normal when it wasn't?

I had no equipment, no sedative with which to dart her, not even a cage in which to carry her, but I didn't hesitate as I ran back to get Josiah, Hudson and the car. Peggy screamed in pain when we picked her up, but she cooperated. She was fully awake and seemed aware of what was happening as I put her in the back of the car and surrounded her with food. I decided to risk the trip to IPR; a two-hour drive, four potatoes, four carrots, half a cabbage, five tomatoes and a loaf of bread later, we made it. Peggy looked at me once or twice when we hit a bump in the road and she was jostled, or between mouthfuls of food. I will never know what she was thinking, but I felt that years of mutual trust were being called on.

At the Institute, we discussed options. They would fix the fracture if they could, and Peggy would be returned to Pumphouse. But if they had to amputate the leg, Peggy would be unable to survive in the wild. They tried everything. Even the people at the Primate Institute, who had seen thousands of baboons, had never encountered anyone like this great old lady. They cleared up the infection and made her comfortable, but in the end there was nothing to be done. Moreover, her teeth told me what I had already guessed: they were so worn down that her days were numbered; even under the best of circumstances, she would soon lose condition and succumb to disease.

There was so much to remember about Peggy: her way of handling infants, and of grooming, which was special and unique to her because she had only one good eye; her calm, almost regal demeanor, in such striking contrast to that of her daughter and granddaughters. What stood out most in my memory was her social skill, her intelligent sophistication: the way she managed to get Sumner to share his meat with her, and in later years, when the males changed and Dr. Bob wasn't so inclined, how she had outmaneuvered this smart male simply by using her wits.

Peggy had given me many crucial insights about baboons. Her life had taught me what was important and what was trivial. But the most meaningful lesson was much more personal: Peggy taught me that you can have strong emotions, such as the special attachment I felt for her, and still do good science. The two were not, as I had once thought, mutually exclusive. In fact, they could be related. I was now involved with crop raiding and rural development, not because they were my most interesting intellectual avenues, but because I wanted to save the baboons. Yet emotions need not overwhelm science. Techniques could still be systematic and rigorous, data could still be safeguarded from bias, interpretations could still be put on a firm quantitative footing. Best of all, feeling strongly about the baboons made the science more rewarding.

Peggy's death was the end of an era. It took place exactly ten years from when I began my first study of Pumphouse. The baboons and I had come a long way, but we still had a long way to go.

The real crisis had arrived. How had it happened, and how could I stop it?

Cripple Troop had discovered the army garbage pits in 1975, seven years ago. They had moved over there more and more until I hardly ever saw them on Kekopey. There were obvious benefits—plenty to eat, even if it was strange human food, and not too far to walk for either food or water. There was even a convenient sleeping site nearby. But there were costs: as the troop's name implied, the high-tension wires had claimed many victims, and the baboons always seemed to have more injuries than was common.

Wabaya led the way to the army camp. The drought had worsened, and with fewer *shambas* to raid, the camp seemed a good alternative for baboons that had developed a preference for the easy life. At first the women and children living in the married quarters felt sorry for the Wabaya mothers and babies, worried that they didn't have enough to eat during these hard times, and offered them tidbits. Also, they enjoyed this close-up look at the baboons. Once you could really see them, they were fascinating, even if you were ignorant of all the details of families, friendships and dominance ranks. There weren't many diversions at the army camp, and the baboons filled the place of a good TV show. Everyone was happy.

But once invited in, the baboons did what baboons do best. They became opportunists, sampling everything in sight—not just handouts but the crops that were starting to emerge from the small irrigated gardens. A few of the larger males, already familiar with army barracks from their childhood in Cripple Troop, were even bolder. They pushed open doors, wandered inside kitchens and storerooms, creating pandemonium and a royal mess.

The love affair soured. The camp residents no longer wanted the baboons around. But the situation wasn't clear-cut: one woman screamed at me to get rid of the baboons, while at the same time her neighbor was handing out food to the troop. The problems we encountered with the Kekopey farmers seemed mild by comparison; at least the farmers had not actively encouraged the monkeys.

It was ironic: We had reached a point with the Kekopey settlers that showed hope for the future; the animals were being actively kept away; our research was pointing in promising new directions and the rural development project had made a good start, perhaps offering a workable solution to the problems of baboons and people coexisting successfully on Kekopey.

Then came the shootings. Fortunately I took action so fast that only one animal was killed and one disappeared. Also fortunately, it was not an official act; an irate sergeant had illegally persuaded the local game ranger to do the shooting. The higher army officials had not sanctioned it, and were receptive to my discussion of the problem's history and its possible solutions. They would not kill any more baboons, but they wanted assurances that the baboon problem would be solved as fast as possible. How I did it was my business. Until then, they would, as they put it, cooperate "in every way we can."

# 15. Searching

By late 1982, I knew we would have to move the Pumphouse Gang; the alternative was the eventual annihilation of the troop. The challenge was clear. In order to solve the army's baboon problem, 131 baboons had to be moved—there were 57 in Pumphouse, 38 in Cripple and 36 in Wabaya—together with their observers, to a region where both could thrive. But where? The national parks were out of the question; there were too many baboons there already. I didn't feel up to the difficult task of negotiating for the use of tribal lands, especially considering the local attitude toward baboons.

This left the gigantic cattle ranches owned by whites. When I'd first come to Kekopey, I'd heard that most white landowners divide primates into three classes. White landowners, of course, were Mark I. Natives were Mark II and baboons were Mark III. It was only in 1976 that baboons had been legally upgraded from the status of vermin to that of wildlife, and even now they were regarded as good for target practice and not much else. Female American baboon researchers were still unclassified, but I had an uncomfortable suspicion as to the category in which I'd fall.

Such bigotry was rumored to be most entrenched in the highlands, including the area around Laikipia, where I hoped to search for our new home. Wealthy Europeans had begun settling the "white" highlands around 1900, and during the wild days of the twenties and thirties engaged in more excesses of drink, drugs, orgies, murders and suicides than seemed possible, given their rudimentary forms of transportation.

My own role models of white landowners were, of course, the Coles. Their attitude toward their native staff was in keeping with their background, and relied heavily on the old idea of patronage. Initially it had been hard for me to accept the disparity between black and white living conditions and material goods, but in 1972 Kekopey's conditions were certainly among the best. Many services were free: medical attention, help with schooling, land for retirement. Whenever anything went wrong, the workers came to the Coles knowing that they would be helped. When I came to Kekopey, liberal Berkeley ideas and all, I saw for the first time how complicated such relationships could be. Contradictions were obvious, but the undercurrent of kindness, respect and concern on both sides could not be ignored.

The Coles were exceptional landowners, certainly, but many of my other contacts were also sympathetic to the rights and causes of native Africans, both human and animal. No matter how colonial in background or aristocratic in taste, many seemed to maintain deep roots of sympathy for the land and its inhabitants. Bwana Cathcart was an example. At least in his seventies, a little stiff with age and bowlegged from many years on a horse, he was still an attractive man, with pale hair either sandy from the sun or white with age—it was hard to tell which. He was the patron of polo in Kenya, and photos of him with English royalty and famous Indian players lined the walls of his polo room. His ranch housed the polo field and the simple structure called the Polo Club. Everyone treated him with respect because of his age, his role and the fact that he was still a fine polo player.

Bwana Cathcart believed in the propriety of human relationships and was greatly upset if a younger player lost his temper on the field or used obscenities. One day as he was being helped to mount his horse, I laughed at something and he mistook my laughter as being directed at him. He admonished me gently, and reminded me that I, too, would probably need help when I reached his age. I assured him that I already did, and we went back to our respective sports—Bwana Cathcart to his polo and I to people watching, for which the Polo Grounds were a good place.

Bwana Cathcart clearly regarded his African servants as being of a different class, yet he treated them in a way that was both just and formal. For me he typified many of the nearby landowners. But there was another, less pleasant type. Philip Turner was a landowner I tried to avoid socially. He was thin and wizened; his withered face and body seemed parched with hatred. Charming, bright and lively within his own social circle, he would become transformed when discussing blacks, the current state of affairs in Kenya and the future. After downing a few drinks, he would spew forth a vicious stream of profanity on all these subjects. His relationship with his native staff was based on fear and threats, though I don't believe he ever abused them physically. It was clear to me that this attitude toward the indigenous Africans carried over into a hatred of the animals I was studying and the very idea of such research. It was as if the baboons were a symbol of everything such people hated and feared. This was brought home to me in a rather frightening way. Kubwa was a ranch near Kekopey, and a grisly discovery was made there: nine baboon skeletons were found in an abandoned water tank. Newson, the assistant manager, insisted that the animals had drowned, which seemed bizarre and unlikely. On further examination, bullet holes and other evidence of human violence were obvious.

In murder mysteries, the one who finds the body is sometimes the most likely suspect. In this instance, the case against Newson was particularly cogent. He was a man in his twenties, though he looked older. His short blond hair stuck up in bristles like an army recruit's, and he waddled when he walked, his gut plunging over his baggy shorts. He swaggered around like a bully and always seemed to be cracking an imaginary whip. He hated baboons, scientists and blacks. Nobody had rights except Newson or those of whom he approved. The workers hated him for his stubborn, radical imperialism; they were not even allowed to leave the ranch's boundaries in their spare time. Consequently no one on the staff worked, except when Newson was actually watching them.

It was easy to see Newson as the destroyer of the nine baboons found in the water tank—even easier when it was discovered that the bullets matched those of his rifle. Strand, the manager of Kubwa, was told about the bodies. He and Newson made a comical pair, Strand being a tall, lean man in his early fifties, shy and fidgety. I had expected Strand's support, because he seemed an intelligent man. But I was wrong.

It was during this crisis that I was trying to find a new home for Pumphouse in Laikipia. One day I received a letter from Strand that said, among other things, "You people plus the baboons are nonproductive and destructive to farming, as are other wild game, trespassers, thieves and illegal *changaa* brewers . . ."

This was not an auspicious start to the "move." Phrases from Strand's vituperative letter kept jumping into my mind as I anticipated taking on the white landowners and their managers. I was spurred by it, too: challenges that might overwhelm me if faced for my own well-being I confronted eagerly when the safety of the animals was at stake. Jonah and I were both sure that somewhere in Laikipia lay the perfect new home.

Baboon paradise would have to be composed of oxymorons. We would need land with a permanent water supply but little rainfall, for rains would mean the arrival of agriculture, with all its attendant problems. Moreover, while the baboons would need ample, varied vegetation and plenty of sleeping cliffs, their observers would require that they be easily visible and on safe terrain. Finally, the area should be isolated enough for the animals to be protected from disturbances, but should also be near a mail drop, roads and human company. Unlike me, some of the Kenyan assistants would be unhappy if the site were too far out in the bush.

We set off in Jonah's plane, TCL—or Tango Charlie Lima, its call name. TCL was a very old Cessna. The single prop filled the cabin with noise, forcing me to wear cotton stuffed in my ears. The wind was directed through the vents—a primitive form of air conditioning—and the cracks in the no longer tight-fitting windows and doors all contributed to the racket. As a result, we were unable to discuss anything we saw in detail until after the flight. This was extraordinarily frustrating, because Jonah's years of flying and of being in the wild had given him an amazing acuity. When we were first together, I thought he was constantly pulling my leg: he would point to a distant hill and identify a tree or an animal; I would refuse to even look at it through my binoculars, because I was sure I was being teased. But time and again Jonah was right: there *was* an augur buzzard perched on that acacia tree on the horizon; there *were* three reedbuck hidden in the dense bush at the base of that mountain.

Binoculars are of little use in a plane. In order to see animals from the air you need to develop a "search image" quite different from viewing the same animals on the ground. Without it you can fly over a multitude of game and simply not see it. On previous flights I had managed to improve my search image, but still came nowhere close to Jonah's skills. He could count animals from a plane in one quick pass-over and be only one or two off in a herd of a hundred buffalo.

We had various scale maps of the area, but here, too, I relied on Jonah. I was terrible at map reading. Flying with Jonah was a serene experience. The exhilaration and adventure depended on the beauty of the sights below one rather than on the derring-do of the pilot. Even when that serenity was disturbed, I always felt safe with him.

As I gazed down on the endless stretches of grassland, I con-gratulated myself on my hard-won ability to recognize at least some animals. My half-trained eyes found the dry season conditions the most difficult, because everything was a dusty straw color into which the many sand-colored ungulates blended easily. When I saw the Thom-son's gazelles from the ground, the dramatic black stripes on their sides were clearly visible. From the air, only a mass of beige backs caught the eye. Even the zebras seemed to fade into the landscape, for the heat haze made their stripes disappear.

Grassland spread in all directions, with only an occasional small acacia tree breaking the sameness; in the small gullies, the vegetation was denser and more varied. For a few seconds, far below us, we sometimes glimpsed a brilliant narrow band of blue water, a seasonal river edged with bright green bushes, its banks lined with giraffe, gazelle and zebra. The only reminder that this grassy wilderness was both cattle country and private land was the occasional well or cattle trough.

I wondered what this landscape would look like after the rains, when the parched ground would turn the incredible new green I had seen each season on Kekopey. As if in answer to my question a patch of green appeared in the midst of the straw. It was an extraordinary oasis: a formal English garden with a perfectly manicured lawn surrounded by roses, bougainvillea and a variety of brilliant flowers. In the center stood an unusual rambling house, much of it obviously made from local materials—stone, cement, wood and glass, with an incongruous tin roof. Large covered verandas with inviting overstuffed armchairs spoke of gracious living. Several horses were corralled nearby, and the entire compound was enclosed with fencing and protective planting. A Land-

Rover and a few flatbed pickups indicated the start of a tiny dirt track that ended at the compound. This track connected with other small ones and then with a large dirt road leading off into the distance.

I imagined the inhabitants hearing and resenting our small plane as it invaded their land and air space; they probably felt secure and isolated, safe from demands and change. But their view was earthbound, while from our vantage point above them we could see that they were neither secure nor isolated. Agriculture was approaching from all directions, eating away part of Laikipia in the south, east and west.

Finally we approached the edge of the plateau, and I gave a yelp that penetrated the noise and Jonah's earphones. We both smiled. The landscape had changed dramatically; there were no more endless plains scattered with animals. We had come to the edge of the world. Small gullies became larger ones, turned into canyons and then into deep gorges as the plateau dropped off into the lowlands below. The place gave the impression of a recently watered Grand Canyon. Except for the dramatic cliffs, the jutting edges were softened with dense, clinging vegetation, and far below I could see a gleaming river.

Here I didn't need binoculars. There were zebra, eland, impala, tommies—and perfect baboon sleeping sites. There was a stream with a pool. And there were baboons! It was reassuring to know that I could recognize baboon country and see it in the same perspective that the animals did. Everywhere I thought there would be monkeys, there they were.

We headed north along the edge of the plateau, watching the high country give way to the dry lowlands, often dramatically in drops of hundreds of feet. In many places the vegetation seemed almost impenetrable, while in others small clearings were crowded with animals feeding on the juicy short grass. It was superb country and it was also baboon country, but in many spots the going would be rough and even dangerous for the baboon watchers. There were many animals, only a small fraction of them visible, and once away from the open areas, visibility was poor. Although wild animals tend to be afraid of humans and to run away, they need to be given ample warning; turning a corner or rounding a bush and running into a human is a harrowing experience for a wild animal, especially one that has been hunted. Species which would actually prefer to run and hide sometimes feel compelled to stand their ground or even attack the intruder.

As we traveled northward, we also tracked the rainfall gradient. Lushness gave way to dry grassland and then to semi-arid country.

Bushes and shrubs were less leafy; everything seemed to be protected by thorns and separated by expanses of bare, dry soil, We turned and headed east. On the northeastern periphery we once again found spectacular gorges and canyons, although the vegetation was not as lush. One gigantic permanent river, the Uaso Ngiro, ran the length of the Laikipia here, branching out into two smaller offshoots. As we moved away from the river, the vegetation to the west remained tangled and profuse, but to the east grassland stretched over a large area dotted with the cindery cones that indicated prehistoric volcanic activity. Here and there were rock outcroppings—kopjes—some only twenty feet high, some magnificently higher. The grassland looked cultivated, almost seeded like a real lawn whose owners had forgotten to water it. This grassy belt was edged by a small mountain range that hid the precipitous drop to the arid lands below.

Now, looking due south, we could see Mount Kenya, standing like a guardian angel at the entrance to this paradise. We circled and headed home; the clouds around the glacial peak of the mountain were dispersing, and the ice reflected the magical pinks and lavenders of the setting sun. I was exhausted, and filled with nostalgia. The physical eyestrain and the spiritual impact of seeing landscape of so much beauty on such a magnificent scale were sensations I had almost forgotten in the last few years, as I watched an embattled Kekopey succumb to the onslaughts of civilization.

I felt my heart sink as I glimpsed the tin roofs of Kekopey below me; I trudged past them daily while I focused on my work, turning my eyes aside and my thoughts inward; once out with the baboons, I could sometimes forget their existence. Looking down from above, I saw the fragility of that small area I called *the field*, saw once again how interwoven it was with human needs and problems. I wondered what Jonah thought; he'd never known Kekopey when it was closer to the unencumbered wilderness he so loved. Yet he had dealt with people throughout his research at Amboseli, including them in his equation, recognizing that the future of wildlife depended on its rational integration with human needs and concerns.

———

I hoped that after I'd met a few of the "nice" ranchers, their estates would seem less threatening. But I didn't know how to meet them. Again, it was Jonah who came up with the answer: Iain Douglas-Hamilton.

Iain was a mythic figure of African wildlife adventure and research, whose remarkable work on elephants was world famous. His non-professional exploits were equally astounding. He was a swashbuckling man with sandy blond hair in a tousled pageboy, whose thick-rimmed glasses glinted above the biggest grin I'd ever seen. He seemed the manifestation of mischief, and one felt one could find anything from a thousand-dollar bill to a live toad in the pockets of his khaki shorts. Hair-raising anecdotes involving Iain dotted the African literary land-scape. Some authors found themselves surrounded by stampeding ele-phants while Iain stood by calmly taking photos and making notes. Other literary lights found themselves *in* the photos—terrified faces in the foreground with charging buffalo, roaring lion or thundering rogue elephant as the backdrop. You could starve on the plains with Iain or nearly die of thirst with him in the bush. You could cling to the door handle as a Land-Rover jumped over fallen trees—or suffer medical or obstetrical calamities miles from civilization and have him save your life.

Or you could fly with him. "What a fantastic little plane!" Iain ran his fingers over the Cessna. Jonah had gone to a conference in Oman and had lent Iain his plane on condition he fly me around and introduce me to some of the landowners he felt might be sympathetic to the Project.

Iain had just crashed his own plane—not his fault, he assured me, though he did add that his friends didn't think so; he had asked for trouble so often that the fact it was mechanical failure this time seemed irrelevant. No one had been injured and "It taught me something," Iain told me. "When you're carrying precious cargo, like your children, you shouldn't take chances."

This was somewhat comforting, but I couldn't help wondering how Iain flew when he *was* taking chances. He seemed entranced with the "fantastic little plane," and kept wondering—aloud—what it would do. We were fated to find out: Iain pushed TCL to the limits of the possible, stalling and banking so steeply that I looked out the window at nothing but the clear October sky, missing trees by centimeters and flirting with whirlwinds of birds.

The positive effect of Iain's pyrotechnics was that they obscured my anxieties about meeting the ranchers. As we landed I had white knuck-les and a prayer on my lips; I was so happy to be on solid ground that the task ahead seemed like child's play.

Iain knew everybody, and gave me stalwart support, even though I don't think he realized how difficult it would be to find a home for Pumphouse, nor how unwelcoming and astonished the ranchers would be. The visits all took on a kind of similarity. I would smile my most charming smile, admire the estate and present my proposal: "We're looking for a place to move about a hundred and thirty baboons."

"Ah, you want to move out some of my baboons! Smashing! I was just getting ready to shoot the bloody vermin."

The look I received as I corrected this impression was always unforgettable, but gradually the ranchers started to listen to me. Iain's presence made all the difference. It was decontaminating: the baboon contact was incompatible with the Douglas-Hamilton contact, and since Iain was with me and the baboons weren't, it was this association that the owners and managers honored.

One of the most promising places proved to be out of the question. It was a 50,000-acre ranch owned by a man I labeled "The Phantom"; he kept popping up in conversations but was never seen. I was told he was young, energetic, and lived on his gigantic estate with no one but his wife for company. Another white man, even a manager, would be too many people. Four baboon researchers would obviously seriously overcrowd the place, and heaven forbid that he should catch the glint of sun off a binocular lens.

I explored two other ranches. The first intrigued me, not simply because I knew there were baboons already there, but because of its owner, Kuki Gallmann. Kuki was a striking-looking Italian woman in her late thirties. She had made Kenya her home for a decade—ten years that had ended in double tragedy when her husband had been killed in a car accident and her teenage son fatally bitten by a puff adder. Kuki dressed with stylish elegance in green khaki set off by exquisite gold jewelry during the day. Her blond hair was swept away from her face, but she had a way of shaking her head at the beginning or end of a thought, almost as if she were tossing back a loose-flowing mane. For all her coiffed stylishness, she loved the bush and fitted in better than some of the traditional Bushmen I have known. She could outdistance most hardy walkers and often, accompanied by an armed guard, roamed for miles tracking her beloved rhinos. Her ranch had one of the largest remaining rhino populations in Kenya. Then she would reappear in the evenings dressed in designer clothes and serving gourmet Italian cuisine as if this was the middle of Tuscany, not miles from the nearest source of provisions and a continent away from the delicacies we were enjoying.

Kuki's heart and soul were dedicated to Sveva, her remaining child, and to the memory of her husband and son and what they loved most: the land and its animals. She had had no formal education in conservation or economics, yet she had managed to think through a plan for the future of her ranch that incorporated some extremely sophisticated ideas. Her land was gorgeous. The Mukutan Gorge cut through the plateau in this region, its steep sides encrusted with vegetation; a clear, deep river wound through the flume below. I glimpsed sleeping cliffs there, and was delighted to spot a troop of about fifty olive baboons *wahooing* at us as we decended. But despite its beauty, the land was unsuitable for our research. Dense leleshwa shrub covered much of it. These bushes, which would eventually grow to tree height, would not only hide the baboons from observers, but rhino, buffalo, elephant and lion as well.

Kuki and her manager understood my plight and that of Pumphouse, and I sensed that they would offer their land as a last resort, a place where the animals could survive if no other site could be found. It would mean the end of the research study, but it might be the only alternative. Kuki's kindness, and that of her manager, gave my spirits an enormous boost.

The second ranch that seemed full of possibilities was Colcheccio. Near enough to Kuki's ranch so that much of it was thick bush, Colcheccio was farther southeast and also had areas of more open plains, some very similar to Kekopey. Iain introduced me to its manager, Stefano Cheli, and to its owner, the count. Kuki had mentioned the count to me; she described him as "large," as if that one word covered all categories, physical, mental, spiritual. "Meet him in Laikipia, not Nairobi," she advised. "If you meet him in Nairobi he'll turn you down in the time it takes to drink a cup of coffee. But he won't say no here—at least not right away."

Iain and I made a bargain: he was still without a plane and needed a way to get to Kuki's for lunch with some visiting European royalty. In return for the use of TCL, he would drop me off at Colcheccio for a meeting with the count.

We landed to the accompaniment of Iain's usual "What a fantastic little plane!" There was no time for him to introduce me to the count, so I went alone, my heart pounding. I hadn't felt this nervous since taking my Ph.D. orals. Poor count! There had been some kind of miscommunication, and he'd been expecting both of us. Why was I alone? I tried to invoke Iain's decontaminating presence, but wasn't surprised when the count merely laughed and said he had too many baboons as it was, hated them and was about to shoot them. I didn't quite believe him. He obviously loved the country and the animals. He'd given up hunting years before, because—he seemed a little sheepish as he said it—times had changed; there were too many hunters and too few animals. He made a grab at his plump midriff and told me he couldn't dive under bushes or climb trees as he had in his youth; besides, he loved seeing the animals and knowing his land was a haven for them. As indeed it was. He didn't have the exact figures, but there was a remarkable number and variety of wildlife. "I wish I had more lions," he said, wistfully. "I'd take Kuki's anytime."

Another clue to the count's character appeared as we were discussing the baboons and the Kenyan habit of using them for target practice. "I want them shot," he said, "but I can't do it. It's like shooting children." It was then that I saw I might stand a chance with him.

By the time we had finished lunch I felt more comfortable, and the count was clearly warming up. We sat in his lodge on the edge of the escarpment and looked across at more than thirty miles of the most beautiful country in Kenya. From our position, not a soul was visible; it was like being totally alone in a vast wilderness. Below us, at the foot of the scarp, kudu and waterbuck were drinking from a spring. As I watched, several baboons came down to drink; the rest of the troop remained hidden in the shade of the bushes and trees spread up the opposite hillside.

By now the count was mumbling something about "seeing what we can do for you," but it was little more than a mumble. Then he asked what the baboons would need. Water, I said, sleeping cliffs and about fifteen square miles to forage over. It was almost—but not quite—in response to this that he had said, "I want to show you a spot."

The "spot" was exquisite. A permanent pool lay half hidden under a row of cliffs dotted with enormous boulders; the cliffs descended to the water's edge, where, like an oasis in the middle of a parched land, two date palms jutted out between the boulders and the ground. The open side was bare, and there were traces of cattle, buffalo and elephants. A tangle of vegetation bordered the perimeter, and I recognized some favorite baboon foods, while there were others I knew they would like.

As we explored Colcheccio, I sensed a change in both our attitudes. Although I was entranced by its beauty, I was worried. The baboons would thrive, and *I* wouldn't care about being isolated in such a remote spot, but the vegetation was thick, and there were enough large and dangerous animals to warrant caution. Moreover, the Kenyan assistants might feel it was too deep in the bush and be unhappy. But in the meantime, I focused all my mind on listening to the count, trying as subtly as I could to rechannel his ideas about baboons. He felt they were responsible for the destruction of his beloved birds, young antelope and small game. I knew that Pumphouse, the most predatory of any baboons observed, had not had a damaging impact on prey populations. In fact, in the early 1970s FAO (the Food and Agriculture Organization), a UN agency, had conducted an experiment that proved this. The organization had wished to determine whether wild animals could be harvested for meat, thus providing an incentive for ranchers to keep wildlife alongside their cattle. In order to determine this, they had to find out the level to which these wildlife populations could be cropped and still rebound in short order. In their Kekopey operations, they removed about half the tommy and impala populations; within three years, and *despite* Pumphouse, the cropped populations were back to their old levels.

By the time Iain returned from his lunch date to pick me up, the count's attitude toward the baboons seemed to have softened; certainly he'd never sized up Colcheccio by looking at it through my baboon spectacles before, and it intrigued him. Toward the end of my tour with him, he'd even been pointing out good sleeping sites.

Softening of the heart is one thing; promising a home is another. Back in the "fantastic little plane" I tried to come to terms with the likelihood that the count would refuse us Colcheccio. His conversation had reminded me of similar ones I'd had with other ranchers, and of comments overheard between tourists in the lodges. Despite their love of wildlife, many of these people shared a view of nature similar to what I'd espoused before I'd come to live with the baboons. Nature existed "out there," untouched by humans, unchanged and unchanging, noble and in perfect equilibrium. In this idyllic realm there were births but no deaths, and populations remained perfectly attuned.

The count was an experienced hunter, so his view of nature was not quite as naïve. He was perfectly aware that death played an important role, and that the natural world was divided into the hunters and the hunted, yet even he had little sense of how ecosystems and communities work, the built-in checks and balances that give each species its own individual part to play. His intention was noble: to preserve his wildlife.

But baboons *were* wildlife, part of a natural ecosystem of predators and prey. Removing even one component of the system upset the balance, often with disastrous consequences. If the count's baboons were increasing in number—as he claimed—it meant that there was still room in their particular niche. When they reached their limit, the population would level off or even decline.

Iain climbed and stalled, banked, grinned and sang. I wondered whether I had convinced the count; he'd looked thoughtful, and his farewell handshake had been a warm one.

The next day, October 11, 1983, I received my answer. The count had changed his mind: Colcheccio was ours if we wanted it.

# 16. Desperation–
# and a Happy Ending

The count's permission seemed a personal triumph, and I was happy as we set off very early that hot Wednesday morning in October 1983. My companions were Josiah and Mary O'Bryan. Mary, who had joined the Baboon Project in September 1983, was a British-American graduate student who had only recently survived a disastrous translocation project of her own. In 1981, she had been assigned by her university adviser to study a deer population that was being moved from Angel Island in northern California, which they had overpopulated. They were devoid of natural enemies, and had eaten themselves out of supplies; the choice of cropping or killing some of them had been greeted with outrage by animal lovers.

Everything about that translocation had gone wrong. Many of the deer, once in their mainland California home, were struck by cars, some were killed by farmers and others starved or simply disappeared. Mary, who had placed radio transmitters on the animals, had to listen to the fading signals as each animal went to its death.

I would have expected her to be tremendously distressed at the idea of another translocation project, but she was even more enthusiastic than I was. Now that she knew what to avoid, she seemed to feel herself safe from further emotional damage and was ready to witness a successful move. Her presence reminded me of how much I was taking the actual move for granted. My main focus had been on an area in which to settle; I had thought of the move as almost a foregone conclusion. Endangered animals have rarely been translocated; they are usually taken into protective custody in captivity. Members of only a few species of birds and ungulates have been moved from one wild place to another, and until now only one primate move of this type has been attempted: a troop of baboons was trapped in Natal and released into a newly created national park previously denuded of its wildlife. The outcome is as yet unknown, since the animals were not followed.

What I wanted to do with the Pumphouse Gang involved a variety of risks: they would be placed, as strangers and intruders, in the home range of other baboon groups which were already familiar with the area and which might strongly object to additional residents. Moreover, Pumphouse had encountered few cat predators on Kekopey, since these had to be controlled to safeguard the cattle. In the new area, the animals would be faced with strange landscapes, strange foods and possibly many predators as well.

Translocation was a last-ditch effort. I felt sure that the animals would have at least a fifty-fifty chance if I did move them, and faced certain death if I did not. I also felt committed to the move as an important experiment that might play a critical role in the future of other primate groups. Pumphouse was scientifically valuable and worth saving for this reason alone. While baboons themselves were not an endangered species, many other primates were. What we would learn from this test of baboons might help save other truly endangered species.

Mary, Josiah and I set off for Colcheccio in an old Suzuki Jeep loaned by a friend—our research car was out of commission. We were quite a sight: I drove, with Mary next to me and Josiah perched on top of our supplies in the back, wedged under a roof intact enough to half smother him but so torn that it offered no protection against the weather. We drove for three hours though some of the most beautiful agricultural land in Kenya, then through increasingly arid country; finally, just as we thought we were lost, we arrived.

Trying to explore unfamiliar territory at ground level when all one has to go on is an aerial view is incredibly difficult. When Iain and I had flown over Colcheccio earlier in October, I had noticed wonderful baboon spots with good sleeping places and one area with a magnificent gorge. The area we found from the air began with a swampy piece of grassland at the top of Colecchio that descended gradually to the Uaso Ngiro River far below. As the descent picked up speed, we saw granite rocks that would make perfect sleeping cliffs, while a trickle of water—the dry-season version of the river—filled small but adequate drinking pools all the way down from the middle region to the bottom of the gorge.

We reached our destination—the Ndebele dam on the Penguin River—just as darkness fell. But the gorge we had seen from the air was more like a gully at ground level. There was plenty of water, but no sleeping sites. We would have to move on.

The next day we looked farther afield. We all tried to stay optimistic, but the more territory we surveyed, the more difficult it became. There were very few sleeping cliffs, even fewer that were close to a permanent water source, and several of these were off limits because the count wanted to ensure that we wouldn't be walking across the beautiful view he had from his lodge. I couldn't blame him. It would have been quite a shock to see several baboon watchers tagging along in the wake of yet another baboon troop.

Finally we located a suitable area on Colcheccio, in the vast open plain of the plateau, rich in grasses and herbs. It was exciting to see so many zebra, tommies and kongoni, and reassuring to find that many of the dangerous animals were either not present or could be seen while at a safe distance. As we drove along, Josiah and I excitedly called out the names of familiar baboon foods as we spotted them; the animals wouldn't starve here, and there was permanent water in several dams. The only problem was that there were no sleeping sites. However, there were plenty of boulders around; why couldn't we *build* a set of sleeping cliffs near one of the dams? The cost would be minimal compared with the other translocation expenses. But after a moment, even I laughed. It would involve building too many artificial cliffs, for baboons change their sleeping sites when one becomes too soiled. They are finicky animals, and prefer clean places for resting.

What were the other possibilities? There were some good stands of fever trees in the swampy area where the river originated. Acacia xanthopholea are quite common in this part of Africa, and baboons in many areas use them as sleeping sites. But could Pumphouse adapt to sleeping in trees after a lifetime of using the cliffs? Their urge to sleep above ground is so strong that they could probably get used to the trees, although I would have preferred to avoid another big adjustment.

We were tired and anxious by the time we returned to the count's lodge, where Stefano, the manager, treated us with hospitality. We had to laugh at our life in the bush: rooms with plenty of hot water and superb Italian cooking. When we needed to discuss our plans, we could sit with a cold drink and watch the magnificent view, complete with greater kudu, waterbuck, giraffe, impala and even the resident baboons.

If we stayed on Colcheccio, we would have to set up our own camp as far from these amenities as possible, and be independent and self-reliant. But Stefano's generosity was so great that I imagined invitations to visit would reach us wherever we were. Nonetheless, I was troubled. Colcheccio was not the baboon paradise I had hoped it would be, either for the animals or for the observers. We decided to postpone further Colcheccio exploration and instead survey one of the other nearby ranches. Stefano offered us the services of Koski, his driver, and we loaded up the Suzuki with four people, a backbreaking and hilarious operation.

At first the new area looked like more of the same. Josiah and Mary had decided that dense bush was our enemy and to be avoided at all costs. This enemy surrounded us completely until we came to Chololo, a neighboring ranch we'd heard about from a friendly farmer. It took us hours to reach it by the necessarily circuitous route we had to follow; it was only a few miles from Colcheccio's southern border, and Jonah and I had flown over it a year earlier. When we finally arrived, Josiah scrambled out of the Jeep and leaped around, shouting, "This is it! This is it! This is it!"

It was. Open plains and rocky outcroppings dotted the landscape. I raced to the top of one: there were permanent dams and temporary water sources. Baboon foods abounded, and the kopje itself was a perfect sleeping site that bore traces of recent occupation. Chololo was on the far eastern part of the huge Laikipia plateau, abutting the Loldaika hills and the escarpment that dropped off to the Samburu lowlands farther east. It was part of an enclave within Laikipia that was grassier and less bushy than most of the area we had traveled through. The baboons would like it and we would, too. While it was far away from any farming and similar enough to Kekopey that adjustment for the monkeys should be easy, it was also sufficiently different that I could test some key scientific questions, as well.

Over the years I had managed to construct a different picture of baboon society based on what I observed and interpreted from the Pumphouse Gang. These animals, although equipped with the same anatomy of aggression as all other baboons, used this potential rarely. Instead they relied upon a complex network of relationships, upon individual social skill and the timely use of sophisticated social maneuvers to achieve their goals. It was important to discover what factors contributed to this type of society, and whether these were unique to Kekopey. Until recently, Kekopey had plenty of baboon foods and few of the baboons' natural predators except humans. Both abundant food and little predator pressure might downplay the role of aggression in baboon life.

Life on Chololo would not be that easy. There was much less food and many more predators. It would be interesting to see if the troop would change its behavior if they moved there or whether, as I suspected, they would find that social strategies were still effective and less costly than employing aggressive methods.

It occurred to me that the baboons and I liked the same type of place; perhaps we simply preferred what we were accustomed to. I could have lived at Colcheccio, but always with some reservations. Like the baboons, I am more at home with open spaces. At that moment I was too excited to eat much of our by now extremely dusty lunch. Josiah seemed more positive than he'd been in a long time. He took the philosophical view that even if we couldn't have Chololo, at least we'd had a wonderful outing. But I had set my heart on the spot, and was feeling better about the future of Pumphouse than I had in a long time.

––––––

We lost Chololo.

It was hard to control my disappointment; we'd been so close to gaining our baboon paradise. I never even got an appointment with the owner; I'd called him to set up one, and he'd extracted more information than I wanted to give him. When he called me back, he said regretfully that he had to refuse our proposal "for confidential reasons." I was stunned. What was he referring to?

Without Chololo, and with Colcheccio as a last resort, we had to look elsewhere. Mary, Josiah and I spent an afternoon studying the map of

Kenya, which had suddenly become an extremely small place. Our requirements forced us to look only in the highlands, areas over 4,000 feet, and everywhere we looked there were people or the threat of people in the near future.

I returned to California for a brief visit in November 1983, and came back to Kenya in December. Mary and Josiah had acted in my place, determinedly taking on the rest of Kenya. First they explored the Masai Mara Game Reserve, the northern part of the most impressive savannah ecosystem in Africa, the Serengeti. There they found vast open plains, wooded rivers, beautiful land with a great deal of wildlife, both permanent residents and Serengeti migrants. But the Mara already had an incredible number of baboons; it might be impossible to slot in over a hundred more, even if we could overcome the bureaucratic red tape. In addition, there were no sleeping cliffs and limited sleeping trees. There were numerous tourist lodges along the rivers, and these would be temptations for the animals. To add to the problems, we discovered that a recent outbreak of tuberculosis had decimated a number of baboon troops.

But even these objections could be overridden; it was politics that made the Mara impossible. Tourists were being robbed and sometimes killed, some claimed, by Tanzanians sneaking over the border. Security was too uncertain for us to consider moving the project there. Other areas—the Kedong, a moonscape fifty miles to the southwest of Nairobi, wild and relatively inaccessible, and Kajiado, a tamer area on the edge of the Athi Plains that border Nairobi to the south—lacked permanent water. It seemed that Laikipia was the only solution.

At this point I decided to venture into an area I had earlier ruled out: tribal lands. The area north of Chololo was Ndorobo and Samburu country. Its borders were similar to Chololo but farther inland it quickly became arid and difficult to traverse. It was then that good fortune struck. There were several privately owned pieces of land in the best part of what I had thought was all tribally owned, and the choicest, a ranch named Mbale, belonged to three brothers of the Samburu tribe. I did have one contact with the tribe: Mike Rainey, a friend of Jonah's, was a biologist, anthropologist, teacher and conservationist whose research on the Samburu spanned more than a decade. His relationship with them was much like Jonah's with the Maasai—mutual respect and trust. I couldn't hope for a better entrée. But Mike was difficult to reach; he lived on the outskirts of distant and rustic Rumuruti, and was constantly away on trips around Kenya. I sent letters and telegrams, and left messages with friends as I watched time slip away.

Mike finally surfaced and agreed to help in reaching the Samburu owners of Mbale. Back into an old Suzuki, this time ours, we went: me, Jonah, Mike and Mike's Samburu assistant, Pakwa. Sempui, the middle Samburu brother, was a stunningly handsome man in his mid-thirties. We all squeezed into our rattletrap vehicle for a tour of Mbale while he talked in Samburu with Mike and Pakwa. Jonah listened with some understanding, as Samburu is a language closely related to Maasai. I couldn't understand what was being said, but as Mike translated for me, and as I watched the kindness in Sempui's face, my spirits rose. I was astonished to discover that Samburu folklore fostered a respect and affection for baboons; in the distant past, so the story goes, some Samburu children went to live with the monkeys and survived a terrible drought, while others perished. The Samburu say that they have baboon mothers.

Our tour of Mbale led us to another wonderful spot for the project. The resources were richer here than in Chololo, even at the end of an unusually dry spell. There was plenty of other wildlife: I spotted two kinds of giraffe; zebra—both the common Burchell variety and the rare, beautiful Grévy's, with their thin stripes and white noses—impala, eland, gerenuk, oryx, all in large numbers and not particularly shy.

We arrived at Sempui's house. Waiting outside under the trees, I listened to melodic voices discussing our fate in a language I couldn't begin to understand. Jonah was my voice to Sempui, speaking in Swahili; Sempui would answer in Samburu. He understood our problem; he, too, felt that baboons and cattle were *not* in conflict, and had no objections to moving the baboons to Mbale; he would discuss the matter with his brothers. He even offered some sound advice about when to move the animals, timing the move with the seasonal rains.

Only two obstacles appeared to block the way. Would his brothers also give their consent to the move? How could they benefit from the project? Should we compensate Sempui for the loss of grazing land when the baboons were there? The monkeys weren't legally ours, and it might set a bad precedent. If we should compensate him, how could we come to an agreement that would be fair?

Maybe helping Mbale with a new project would be the answer. Sempui badly wanted to build a new, deep dam, which seemed to me a possibility, but not a complete solution. What would his brothers require? We couldn't guarantee that the baboons would stay on Sempui's part of the ranch, nor even on Mbale itself. How would the neighbors react?

We parted, resolving to meet again in two weeks. In the meantime, Sempui would discuss the matter with his brothers and the neighbors, and I would try to come up with some negotiating points.

We met exactly fourteen days later. Again, the goats wandered about and the sun beat down while melodic voices rose and fell. But something was wrong. Sempui wouldn't meet my eyes and the brothers had not shown up. Sempui's whole body communicated his discomfort.

I could read on Jonah's face that the problem was a major one. Finally he explained it to me: Sempui was worried that the baboons would begin to kill the kids and lambs, not only on Mbale but throughout the whole area. "Nothing is as bad as coming between a Samburu and his animals," Jonah quoted Sempui. I asked how many small stock *had* been killed by baboons locally. Only two, Sempui replied, in the last two years. But ours would be *different* baboons, he said. Who knew what they might do?

The more the men talked, the clearer it became that some other issue was involved. At last Jonah, Josiah, Mike, Pakwa and I regrouped to decide what to do. Jonah told us that he felt it was hopeless to continue.

What could have happened to change Sempui's mind? At lunch, Mike and Pakwa discussed it with us. They were shocked; each felt he understood the Samburu culture and Sempui himself, and this reversal didn't fit. Our only clue came from an observation of Josiah's. On their return visit, he and Mary had been with Sempui for hours, touring Mbale. At one point Sempui had left them to pay a two-hour visit to the local councillor, the focus of county politics. When he rejoined them, Sempui began to ask probing questions, looking somewhat doubtful. Mary hadn't even noticed the change, but Josiah had. It was possible that the councillor had dissuaded Sempui, either seeing no profit for himself or worrying about the political impact a baboon project could have for the area. There may have been some kind of competition between the councillor and Sempui. We didn't know what was wrong; the point was that Mbale was now out of the question.

We raced back to Chololo to look for the manager, Sammy Jessel, whom Josiah and I had met on an earlier trip. I needed all the composure I could muster to talk with him, and was grateful for the support of my companions. Josiah and Pakwa sat chatting under a tree, while Mike and Jonah stood by as I buttonholed Sammy. I described the history of the project and my ideas for the move as calmly as I could, and was thankful when he graciously took over the conversation, letting me know that he understood the problems. He seemed to comprehend my own position, the place of baboons on his ranch, the objections of the neighbors and even the needs of the animals themselves. His response was thoughtful—not positive exactly, but not abrupt, disbelieving or sarcastic. He promised to raise the issue with his father, John, who was the main director and chairman of the company that owned Chololo.

I didn't have much hope, and the flight back that February day was sad and silent. I tried to be cheerful; at least we had Colcheccio, a fine place for the baboons to live, even if the project would have to come to an end.

Breaking the news to the project workers at Gilgil was the hardest of all. They had expected a triumphant return, but saw at once that the news was bad. I fought to find some enthusiasm and hope to give to the staff. The Kenyans I've known have always tried to make the best of everything; they seldom reveal exactly how they feel, yet as they filed into the meeting room I could see changes. Hudson had lost the lilt in his walk; Simon's smile was not quite so broad as usual; the twinkle in Francis's eyes was dulled. Josiah's face showed only resignation.

I reviewed the events and the course I felt we had to follow. We didn't know what the future held. The baboons might have to go to Colcheccio, but that would be better than having them decimated at Kekopey, and I felt it held some promise for the continuation of the project. After further exploration, perhaps other more observer-friendly sites could be found in Colcheccio.

I explained that we would start with a plan for a two-year follow-up of the move to Colcheccio. I explained that life there would be heard because we would be truly isolated and observation would be difficult and probably dangerous. If the project did come to an end, I pledged to find them all other jobs. I sensed that they had a commitment to the project beyond mere employment. We had become a team; we *were* the Baboon Project.

I tried not to think of Chololo during the next week, assuring myself that Sammy's conversation with his father could only result in a polite refusal. Fortunately it was a busy week; there was not much time for brooding. When I returned to Nairobi there was a message for me to call John Jessel. My heart sank: the call had come too quickly and must surely signal bad news.

I was wrong. John Jessel wanted more information. He pointed out that the baboons would have to pay their way somehow. I couldn't promise large sums of money, but Jonah had helped me work out the basis for a plan when we were still hopeful about Mbale. The same scheme, I told Jessel, could be used at Chololo. Also, I could help to develop tourism there, assisting the efforts already begun by the directors. Jessel listened; he seemed kindly and concerned. What was more, he seemed interested.

I kept as calm as I could, trying not to be too hopeful. Twenty-four hours lasted a week; then the call came, and the answer was yes. I rushed over to Jonah and kissed him in front of the entire office staff, which I suspect shocked them far more than Jessel's acceptance of the project.

Mary and I drove to Nanyuki to meet Jessel and work out the final arrangements. After all was settled, I felt delighted and relieved. I had finally done something for the baboons. They'd needed a new home and I'd found it for them. Their future was no longer soley my responsibility: the ultimate success or failure of the translocation would rest with the monkeys.

It remained for Jonah to shock me with a smug smile. "You've done the impossible; you've found a new home for the baboons," he said. Through all those weeks and months, as he flew me around, translating, negotiating, advising, figuring, introducing and manipulating, I'd relied almost completely on his calm certainty that all would be well. And all the time he'd had grave doubts. In retrospect, it did seem utterly impossible. I was glad he hadn't told me before, glad that he had done so now. There was only one thing left to do: marry him!

So I did just that, here in Kenya. We had hoped to be married in a beautiful place under a beautiful tree, in a quiet, simple ceremony. This was more difficult than it sounds. First, since we didn't want a religious ceremony, we had to find a civil "marriage officer" to perform it. Such marriages are normally performed in government offices, which are dingy and charmless; it took a special dispensation and a great deal of talking to get the marriage officer to agree to come out into the wilds for the ceremony. We were helped by Philip Leakey, both in sorting out the red tape and in finding a spot. Philip was and still is the only white MP in the Kenya government, and an old friend of Jonah's. When we told him of our difficulties—we needed somewhere accessible enough for Philip's pregnant wife, Valerie, and Jonah's seventy-year-old mother to join us—the Leakeys volunteered an exquisite place that formed part of fifteen acres of land they owned on the edge of

Nairobi National Park. It was on the Mbaggathi River, at the top of a rocky gorge, with a magnificent view of the park and its array of wildlife.

The ceremony was performed with only eleven people present, including Jonah's mother and brother, who had flown out from London. The marriage officer said his little piece; the only part I can remember is that he admonished us not to commit bigamy, since we were being married under civil, not tribal law (under tribal law you can have more than one wife). We then exchanged a few words with each other and asked Philip, in his capacity as government official, to give us his blessing. Refreshments were served, including the traditional goat required by Maasai custom, and everyone admired the view. (Three years later we built our own house adjacent to that beautiful spot, on fifteen acres Jonah had bought.)

Our wedding reception was held that evening at a restaurant called The Carnivore. We had chosen it both for its informality and because it specialized in the barbecued meats Jonah's Maasai friends enjoyed. The 125 guests—Maasai, baboon watchers and others—all danced together, the Maasai rubbing shoulders with staid British colonial types and everyone thoroughly enjoying themselves. We went back to work the next day; we'd taken our "busman's honeymoon" months earlier, traveling through the national parks of Indonesia and Malaysia.

The baboons had a future, the Baboon Project had a future and I had a future.

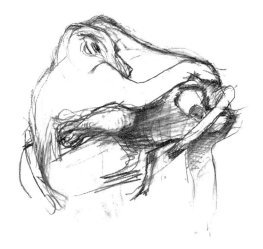

# 17. Capture and Release

In August 1984, I was pregnant and anxious, not for myself but for the baboons. Finding them a new home was only a small part of the problem; we also had to get them there. In 1978–79, when the future of the baboons began looking grim, I had embarked on a project to collect critical biological information from the Kekopey troops we had been studying closely and also from as many other troops as we could. The animals were released as soon as blood and other samples were taken, and no animal was held for more than a day. The total number of baboons captured and examined at that time was close to five hundred, selected from six of the troops. I did have some experience in translocation, but it was on a small scale, with Chumley and Higgins, raiders of the herders' huts. Getting the Pumphouse Gang moved was most important for me, but two other troops were in equal danger from all the changes taking place at Kekopey and more of a problem at the army camp: Cripple Troop, home of the delinquent males of Pumphouse, and Wabaya, the small splinter troop that had taken to crop raiding and the easy life.

I would have liked to move all 131 baboons to Chololo, but there wasn't room. Chololo already had resident troops, and from our surveys we guessed that Pumphouse and Wabaya could squeeze in, but no more. I would move Cripple Troop to Colcheccio. Cripple had not been closely observed for long, so there was less need for good field-watching conditions. I could not only ensure them a safe home, but could also use them as guinea pigs for the more important move of Wabaya and Pumphouse.

The first step was to capture the baboons. The capture team consisted of the Baboon Project members, the staff from the Institute of Primate Research in Nairobi, and, as time went on, a great many other people: local farmers, friends from all over Kenya and help from Ker and Downey, the famous safari company. Bob Campbell came along to film the events for Survival Anglia Television.

Major capture operations in the fifties and sixties had obtained hundreds of animals for medical research in Europe and America, so some techniques were already available to us. But professional trappers had managed to capture only part of any given troop, and since we needed to capture three entire troops, we had to improve on their methods. Moreover, we were concerned not only for the physical welfare of the baboons, but for their psyches.

Our first task was to accustom the animals to the traps. The sites chosen had to fulfill a number of requirements: they had to be out of the way of people living in or traveling through the area in order to avoid accidents and vandalism; the traps themselves had to be on level ground, accessible to the rugged vehicles needed to remove the animals from the site; we also had to be sure the sites we chose would attract only the three troops—Cripple, Wabaya and Pumphouse—out of the ten that used the general area.

The traps were five-foot-high wire-mesh cages with reinforced edges. High inside was a small platform on which bait would be placed: a string was tied to an ear of maize on the platform, threaded through a hole and guided around the top of the cage, where it was attached to the sliding door, keeping it in its upright position. In order to spring the trap, the baboon would have to get completely inside the cage to reach the maize and pull it away from the platform, breaking the string that caused the door to fall. Our aim was to get the baboons used to going into the unset traps for food, to teach them to regard the cages without suspicion and accept them as part of the natural environment.

Once again the place of the army camp in baboon life posed a problem. The baboons living on army garbage already had food sources more palatable than what we were offering them as a lure. Fortunately, the troops themselves provided a solution. Cripple Troop was consistently bullied by Wabaya at the dump; no matter how early Cripple got there, Wabaya would come charging down and chase them away—right to our seductively baited but as yet unset traps.

Within three weeks we had all three troops rushing to the traps to get a treat; they had become so accustomed to the cages that babies used them as jungle gyms, consort couples liked to copulate in them and big males found they provided comfortable footrests. We were ready.

To ensure that we trapped only Cripple and not the bullies in Wabaya, we had chased the latter to a distant sleeping site and appeased them with bananas, pineapples, carrots and cabbage. I didn't sleep the night we set the traps. It took two hours to set the forty cages. Back at the Red House I tried to wait calmly, but ended up arriving at the site well before dawn. Suddenly I heard the monkeys. *Clank!* The first trapdoor falling took me by surprise; I had never really thought it would work. *Clank!* You would think that after one *clank* the monkeys would shy away, but greed overcame suspicion. *Clankclankclankclankclank* . . . within ten minutes the whole troop was captive—except for three brown infants which had entered with their mothers, but had run out before the trap was sprung. My plan had been to sedate the infants, but before I had time to assess the situation the institute team decided that the infants were small enough simply to chase down.

The three babies huddled together at the edge of the trapping area, within easy reach of the safety of the cliffs. If we removed the rest of the troop, the infants would certainly die; if they didn't succumb to predators or the cold nights, they were likely to be so upset that they wouldn't eat properly. While we chased the babies, the trapped animals became more and more restless. We had to begin sedating and removing the baboons. The IPR staff arrived to do this. Normally it would be a quick and relatively painless process, but the pole syringe wasn't working properly. Sometimes it bent, sometimes it delivered only a little of the drug or none at all. I stood on my lookout hill above the site directing operations; with each misfire, my muscles tightened. We finally decided to sedate the monkeys at close range with the blowgun. The sedation darts were difficult to get and valuable, but it was important to remove the animals from their cages as fast as possible; the sun was beginning to get hot, and overheating would be dangerous to the sedated baboons. The IPR staff opened the cages and carried the unconscious baboons to a waiting pickup truck. They looked peaceful, lined up in rows; the awake, alert babies, who had been carefully tied to their mothers with pieces of string for safekeeping until they reached the holding cages, glanced about curiously, clinging desperately to warm, comforting bellies.

About half of the thirty-three trapped animals had been sedated when a big male got loose. Arnold had been trapped with the others, and had sat in his cage quietly. Now he was out, and it was my fault. Earlier, Josiah had suggested that we wire shut the doors of the traps that held males. I had resisted, wanting to avoid unnecessary disruption; while the trapped monkeys were generally quite passive, they tended to panic at the approach of human beings. Arnold had used the time between trapping and sedation to figure out how to get out of the trap. Most of the other animals had accepted their fate calmly. Arnold had been the only one to discover how to lift the heavy trapdoor. Once free, he ran to the other traps and sat looking at the unsprung ones. The last thing I would have thought a sensible baboon would have done would be to try the traps again, but Arnold did just that. He would enter a trap partway, gingerly pick up the maize, sense the tension on the string and jump out. To Arnold, a tight bait meant a set trap: he was not about to make that mistake again. He did the best he could in a bad situation; he began stealing half-eaten cobs from his trapped comrades.

It was obvious that we would never trap Arnold; our only hope was to dart him. We could try either the crossbow, whose greater force allowed us to shoot farther, or the blowgun, which was safe but required close proximity to the animal. We could also try the capture pistol, which was a compromise between the two.

The crossbow seemed the best bet. Joseph Tisot, chief of the IPR team and a crack shot, took aim. The shot fell short. From then on, Arnold sat a few steps beyond what he correctly judged the range of the arrow to be and watched the darts fall at his feet. Our final ruse was to try to use his troop loyalty against him. During the sedation he had rushed to defend any female or infant friend we were trying to approach. Now, using one of his friends as a lure, we tried to get close to him with the blowgun. Arnold glimpsed the gun, detoured in a flash and stubbornly maintained his distance. At this point we gave up and turned our attention back to the three infants.

Joseph had a brilliant idea. If we sedated the infants' mothers and then opened their cage doors, attaching a string so that we could spring the doors by hand, we might persuade the infants to enter the cages. The first mother was sedated, and fell asleep on the floor of her trap. Most of the troop had now been transferred to the pickup, and the relative quiet and calm reassured the infants. They ventured closer; one infant spotted its mother and saw the open door.

In a flight for freedom—so it thought—the baby dashed in to its mother, immediately searching for her nipple and much-needed comfort. The second and third babies were easily captured and I heaved a sigh of relief for the first time since dawn. The last of the baboons was loaded and the pickup drove away to the holding area. I stayed behind to watch Arnold, wondering if he might still yield to the temptation of the maize in the set traps. Soon my frustration at him turned to pity. He seemed a pathetic figure, searching among the traps for his missing troop, rushing to the cliffline to check if they had gone in that direction, then rushing back to the traps again, uttering the desolate *wahoo*, the "lost" call of an unhappy baboon.

––––––––––

Arnold kept searching for two and a half days; he refused to enter a trap and eventually we had to remove the cages to prepare for the capture of Wabaya. I had left him after a couple of hours and rushed to the holding area. We had set up a makeshift lab at the far end of Kekopey to take advantage of our chance to get biological information on individual wild animals unaltered by life in captivity. It was also *my* chance to get real facts about the baboons I knew so well. It was one thing to guess from their appearance that Hoppy was bigger than Kit—two Pumphouse members now residing in Cripple Troop—but it was another to know how much each actually weighed.

Getting such information was of enormous assistance in unraveling the world of baboons. But we were able to go still further: we could look inside mouths, examine crooked fingers, measure subcutaneous fat and testicle descent, check female reproductive condition and generally invade the privacy of every quietly sleeping baboon. We could take fecal samples that would reveal parasite loads and draw blood for tests that could provide information on kinship, perhaps even on paternity.

Aside from getting the technical information, it was thrilling for me to be able to touch and handle the baboons while they were sedated. I had exerted so much willpower over so many years *not* to touch them, *not* to give in to the urge to reach out and make contact that to do so now was surprisingly difficult. Once the first tentative caress had been accomplished, I indulged in an orgy of tweaking, stroking, jiggling. I was surprised by much of what I saw as the animals were worked over. They were in good condition, but even the heaviest had little body fat. Minor problems were common, including cataracts and missing or broken fingers, though there were surprisingly few of the injuries caused by high-tension wires that had given Cripple Troop its name. Mouths were worst. There was not a cavity in the entire troop, but many of the adult males showed severe problems ranging from recent or healed abscesses, especially around the canines, to infections and broken teeth. I became curious about the other troops; it was possible that a diet of army food had caused the damage in the case of Cripple Troop.

Although their mouths were a mess, the baboons were incredibly clean. Hunters and trappers have always remarked that baboons are notably free of parasites compared with other wild animals living in the same environment. Only one small orphan in Cripple Troop showed signs of lice or fleas, reconfirming that social grooming is an effective method of parasite control.

———

Once the animals had been processed by the IPR team, they were watered and fed and left alone as much as possible. I hated seeing them there, three rows of small cages facing each other, protected from the sun by a rickety awning and from the wind and rain by tarps.

Our guinea pigs! From Cripple Troop we learned to make life better for Wabaya and ultimately for Pumphouse. We did find that the cages were too small for the adult males to find comfortable resting positions. Other lessons: Feeding baboons large quantities of cabbage gives them diarrhea; nursing mothers have voracious appetites—one of them grabbed every banana I offered her, and when she ran out of space in her hands she began stuffing them under her armpits. The nursing infants presented a serious problem. Nursing and weaning normally flows smoothly and easily, following their own rules, but under conditions of constant contact and lowered maternal tolerance, weaning becomes violent. Cripple Troop mothers attacked their babies; as a result, the babies became even more insecure and sought further excess contact, beginning a vicious cycle.

The solution seemed to be food. Whenever a mother began attacking

1

B Campbell/National Geographic

1. Translocation: babies were
   tied to their sedated mothers

2. I gave some animals radio
   collars

2

S.C.S.

S.C.S.

3. Sedated monkeys on way to holding area

4. Captive males were an attraction to the rest of the troop

5. Pumphouse's first drink after release

6. Exploring a house near the crops

S.C.S.

S.C.S./National Geographic

4

5

6

S.C.S.

7

S.C.S.

7. Being social: closeness and grooming

8. The baboons' new home at Chololo

9. Brother and sister resting

S.C.S.

S.C.S.

10

S.C.S.

11

10. The baboons found familiar food the first day

11. The house at Chololo

12. Prickly-pear cactus—always a favorite

S.C.S./National Geographic

S.C.S.

13

13. A rest in the shade

her infant, we gave her more food, which occupied her enough so that she allowed her infant to return to suckling; afterward, she would appear less frustrated. In fact, <u>food became our mainstay in helping all the animals survive the boredom and anxiety of captivity.</u> We portioned out helpings of loose dry maize all day long. I found that if I dropped it outside each cage, the animal would be forced to pick up each kernel individually, and finding ways in which to pick up the more distant tidbits became a particularly time- and thought-consuming operation.

The most poignant lesson—and one I should have predicted—concerned social interactions. Cripple Troop members had been placed in their cage locations at random. Mothers kept their young infants with them, but otherwise no attention was paid to who was placed next to whom. The stresses of captivity were thus aggravated by social stresses imposed by unfriendly neighbors. One adolescent male harassed his next-door neighbor so fiercely that I decided to move some cages around. Unfortunately this upset all the baboons: the animals being moved felt helpless and apprehensive, and expressed this by screaming violently. <u>I would have the cages for the other troops placed so that the animals' relationships were taken into account.</u>

---

It was finally time to move the troop to Colcheccio. I had worried about having to stack the cages, but it turned out that the animals in the middle enjoyed the security of being surrounded by the others. We formed a five-vehicle convoy, with me in the front Land-Rover driven by Bob Campbell, cameras in tow, followed by the baboons' truck and three more cars containing the Baboon Project people and the remainder of the IPR staff, who would help with the release.

I wanted to have the truck containing the baboons entirely covered, to be certain they couldn't see where they were going; perhaps, like dogs or cats, they could retrace their steps many miles to return home. But the monkeys needed some ventilation to protect them from the diesel fumes the truck spouted continually, so I placed the tarps as best I could to hide the view and mask the fumes.

Both the truck and its driver were on loan from a friend, Kirti Morjaria. The driver, Owino, was wonderful. He could cope with anything, and endeared himself to me through his sincere concern for the monkeys. I don't think there is a steadier or more careful driver in all Kenya.

It was a cold morning, and an hour into the six-hour trip we stopped so that I could check on the monkeys and let them warm up a little. An amazing sight and sound greeted me as I approached them. As soon as the truck had stopped, the monkeys gave a lengthy series of grunts and began to feed on the remainder of the food in their cages and reach between cages to groom. It reminded me of how adaptable they were as long as their social ties were respected.

Once we'd reached Colcheccio, it took us an hour to travel the remaining eight miles to the release site. The road was clearly marked, but Colcheccio was dense and the landscape rough. At last we pulled up near a bend in the river. The water was visible, there was a cluster of acacia trees growing nearby that was large enough to hold an entire baboon troop and sleeping cliffs were visible to the northwest.

The IPR staff unloaded the cages, and we placed them in a single line facing the river, hoping that the animals would remember where they could find water after their release. Once out of the cages, they could, of course, go anywhere, but I wanted to induce them to stay in the area we had chosen. Despite his generosity in offering us Colecchio, the count wouldn't be happy if Cripple Troop made his lodge and beautiful view their favorite location. This site was the farthest we could get from the lodge. It seemed unlikely that the troop would march over dense, dangerous and unknown territory to end up there. However, other ranches where the monkeys definitely would *not* be welcomed were even closer, so we could still be faced with problems.

We would not release the large males immediately, but hold them captive at the site for several days. I felt that doing this might constrain the rest of the troop so that by the time we did release the males, the main group of females and youngsters would already have established the beginnings of a safe haven. When the males were finally released, the rest of the troop would act as a restraint on their greater tendency to wander.

We would continue to provide the released animals with food for a while; this would help focus their attention on the area as well as supplement their diet until they located suitable natural foods. We placed the males off to one side, the females and youngsters facing the river, and fed and watered them. I hoped to wait until late afternoon to release them, thus giving the animals little time to wander far before nightfall, but as I looked around I realized this wasn't going to be possible. By now the baboons were so distraught that we would have to release them at once before they hurt themselves. Four carried radio-equipped collars, so probably we could find them again if they did wander.

The IPR staff who were to release the baboons climbed on top of the cages after they had unwired the tops of the doors. I gave the signal, and eight animals were freed simultaneously. We had chosen to free the most agitated ones first, and these represented no coherent social or age cluster. Moreover, the sudden release in the midst of so many people terrorized the animals. We had hoped they would rush to the nearby acacia grove to wait for the rest of the troop and to survey the area, but they rapidly placed as much distance between themselves and us as they could manage, and soon disappeared into the dense vegetation. The second group was released and these too hurled themselves into freedom. Everything was happening too quickly.

The last animals were about to be freed when suddenly a female reversed direction and slowly approached the cages. She circled the males to look at the long row of mostly empty transport cages that had held the females and youngsters. She looked warily up and down the rows, glancing at the people around. I signaled everyone to back off, and in that moment she rushed to the cage adjacent to the one she'd been in and grabbed an uneaten ear of maize, then catapulted herself down the riverbank and disappeared in the direction of the retreating troop. Apparently she'd had her eye on the cob for some time, and decided that if she were going out into the unknown she'd better retrieve the tasty morsel.

The sight and sound of the captive males was heartrending. They grunted, first softly, then urgently as their friends disappeared, and strained to get out of the cages. Frustrated, they attacked the cages, turned in circles and stared at the retreating troop. Had I not been so convinced about the importance of keeping the males for a few days, I would not have been able to stand their distress. The sight of a little infant baboon, moaning pitifully during weaning, its body racked convulsively with silent sobs, had moved me many times in the past. The behavior of these males was even more touching as they compromised their normal dignity under what was, for them, an even direr weaning—the loss of the entire troop.

The Cripple Troop females and youngsters had been released around four in the afternoon, and it took Bob Campbell and me more than an hour to catch up with them. They had made a wide arc away from the river and were heading for some distant cliffs, higher and more vertical than the ones we'd chosen for them, and offering greater safety. I was not sure how they would react to us, so we kept a safe distance away. All we could tell was that a few animals had lagged behind; they were up in large bushes looking for the rest of the troop, which by now was at the foothills of the cliffs. The two subgroups exchanged "contact" and "lost" calls for about an hour, trying to decide who was going to make the final decision of where they should go. This helped us account for everyone, and the stalemate between the two groups used up time that otherwise could have taken them far away in an undesired direction.

Finally they made up their minds. The recalcitrant rear went to join the main group, and in silent formation, very unlike their casual movements at Gilgil, proceeded to the sleeping cliffs. Bob and I returned to camp, then I headed off on my own, saying I wanted to check on the males. I could no longer contain myself, and collapsed in front of the male cages, where I relaxed for the first time in weeks. There still might be problems, even failures, but we had come this far with remarkable success. I had not allowed myself to think much about the risks, but below the surface I had lived in constant fear that the baboons would die or disappear, and that I would be responsible. If you have to play God, it's good to *be* God, all-knowing and all-powerful. I was neither, and now was only too happy to let Cripple Troop be the guardians of their own future.

––––––––

The next morning, Mary and Hudson went to find the monkeys with the radio tracking equipment, and Josiah and I stayed behind with the males. They had fallen silent since losing sight of the troop the day before, except for an occasional burst of aggressive energy directed at their cages.

At about nine, the males began to call, making grunts and *wahoos*, and half an hour later a strange troop appeared from the ridge to the east, heading toward the river. Several males from this troop approached our own imprisoned males and climbed a nearby tree, curious and wary. First Benjy, then Hoppy and then all the males produced a passionate series of "contact" calls, staring hard at the indigenous troop. When our captives got no positive response to their calls, they slowly quieted down. Four hours later, the strange troop continued on its way. I was fascinated by the interaction that had taken place, and pleased that it hadn't been hostile.

Mary and Hudson returned with disturbing news: they couldn't locate the troop. Although they had reached the sleeping cliffs early, the monkeys had left. The radio tracking antenna gave no signal at all from the area. All we could do was to continue searching. We divided into teams for safety, going in different directions in order to extend the area covered.

It took us until midafternoon to find the troop. They had moved only a few miles downstream from the release site, returning from drinking in the river and then heading in tight formation back to the ridge below the sleeping cliffs. We decided to carry food to the sleeping cliffs before dawn and provision the troop as they came down the next morning. We would also use teams of followers in relay fashion so as not to lose the troop again.

---

At dawn, Josiah, Mary and I could hear the troop before we saw them. Reaching an open spot, we scattered cabbage, loose maize and maize cobs. The troop was moving in tight single file, led by the young adolescent male Richard (Robin's first child), who had apparently convinced them he was old enough to fill the bill as leader. By now they were moving so fast that I worried they might not notice the food, so Mary and I created a shower of cabbage and maize around us.

It worked. Richard stood up, took about two seconds to assess the situation and was eating furiously before his followers knew what had happened. They joined him and settled down to a frenzy of eating; they were obviously hungry, and also nervous—the slightest sound sent them rushing for cover. They were animals in conflict; the food was a big incentive, but being in the open in this strange new place distressed them. Watching the animals, I saw that Richard's leadership was opportunistic: he would rush to put himself at the head of the troop once it had decided on its direction, but now that it was confused, he was forced to run hither and yon to keep himself in the forefront. When he tried to exert his influence and select a direction, he became a leader with no followers. He finally gave up and returned to the provisioning site to join the rest of the troop in finishing the food.

Then I witnessed a remarkable sight. First one female, then another, then some youngsters moved through the resting cluster of animals, embracing other troop members and finally sitting and grooming with a friend or relative. Embracing is a gesture of greeting, normally exchanged by individuals who are not on close terms; it is a positive statement, a friendly "hello." Cripple Troop's behavior was unprecedented in my experience. Embraces are important but infrequent, yet here was most of the troop clutching one another in greeting as if they were attending a family reunion. It impressed me that perhaps this was the first time the members of Cripple Troop had been able to reestablish the social web disrupted by the events of the week, the network of relationships that had become disoriented by putting the wrong animals next to one another in the holding area and by the trauma of the move. Up until now, they had been too disturbed to resolve the tension; now, finally, normalcy was returning.

I watched, enthralled, as full-bellied baboons sunned, rested or groomed. The severe and aggressive weaning that had emerged during captivity was gone, as were the frustration and signs of depression. The troop even appeared more peaceful and less aggressive than at Gilgil. Although the vegetation and topography of the area was different from Kekopey, the baboons looked as if they belonged there. It all seemed natural and right. Of course, death, injury and disease would not be excluded from their future, but these would be part of a cycle whose agents would not now be so directly human.

But there was a lot left to discover, by both the baboons and me, before either of us could be certain that the future was truly rosy. The monkeys had to learn what there was to eat in their new home, and we had to free the males, hoping that the reunited troop would not wander into areas where they were unwanted.

Progress was made in both directions as the troop established a limited day range that encompassed the "White" sleeping cliffs, past the open area where we provisioned them, down to the river within sight and sound of the males. Although we were still providing a significant amount of their food, they immediately began to feed on local plants and insects. Unfortunately, the drought had devastated the area and there was little to eat; the youngsters, innovative as always, began sampling new foods, rejecting some and munching contentedly on others.

It seemed unproductive to hold the males captive any longer, so we plotted their escape. The troop was was moving down from the sleeping cliffs in the direction of the camp and the males. We decided to wait until the troop could see the captives clearly and vice versa; then, on a signal I would give from high up on the ridge, the males would be set free.

At least, that was the plan. But someone misunderstood. The troop was slow about actually appearing in the right spot, and the males were released before they could see where the troop was. They casually walked away from the cages in a different direction, so we would have to wait and see if, when and how the two halves of the troop connected. In the late afternoon, the released males caught up with the rest of the troop; the translocation was now complete.

Content, the next day I left to resume preparations for capturing and translocating Wabaya. Cripple Troop was feeding and moving normally, although they took fright easily. The other troops were more important and would pose their own problems, but I now had many of the answers I needed. An entire troop *could* be moved; we *could* influence where they went and limit them in ways that were important. Earlier, watching Cripple Troop in their holding cages, I had one pressing question to ask them. Now I had their answer, as surely as if they had spoken to me: they preferred to be free in a strange land rather than secure but captive. Best of all, they seemed to bear no grudges; they would allow me to be their companion in this new world.

# 18. Final Moves

By September 1984, we had learned a great deal about translocation and the Cripple Troop move had been so successful that I
thought moving Wabaya would be easy. These *were* the "bad
guys," the crop raiders who, in alienating the army, had brought the
whole translocation issue to a head. Their dependence on army food
posed a special problem in capturing them. Even though we lured
them with pineapples and bananas as well as cabbage and maize, the
animals refused to go into the traps. Eventually Josiah and Hudson
suggested we collect the army garbage and add it to the bait provided.
Over the years these two had had to perform a number of unusual
tasks in the name of baboon research, but this was a first. The idea
worked.

Finally, all was ready. I arrived at the Wabaya trapping site on
Kekopey before dawn, checked the traps to make certain everything
was in order and then settled down to wait. The troop had slept at

Fig Tree, four miles away, but they were expected to arrive by sunrise. Josiah reported that the troop was slowly moving in the right direction; as kings of the garbage, and with Cripple Troop no longer around to bother with, an hour or two's delay would make little difference.

But nothing in the capture of Wabaya went as smoothly as it had for Cripple Troop. Subgroups arrived at different times, and once the first batch of animals had been captured, the others were much more wary. By the time the vehicle left with the first animals, the sun had heated the uncovered metal to an almost unbearable temperature. No one was able to relax until all the troop were safely in holding cages under a makeshift awning, with plenty of water.

This time we made sure the monkeys were properly arranged; individuals were surrounded by family and friends; males were placed between friends and separated from one another, or from those to whom they might present a threat, by as many animals as possible. The effect was marked; Wabaya was clearly less upset by captivity than Cripple Troop. They were also in much better condition: monopolizing the garbage obviously had real advantages. But male mouths were a mess.

Loading and transporting went as planned. We had to use two trucks, because we were conveying more supplies and additional equipment to Chololo. We divided up the responsibilities; we also banished all but the Baboon Project staff from the actual release. Once the animals were unloaded, watered and fed, we selected those to be released first and then waited. A little before the scheduled time, the baboons became agitated, just as the Cripple Troop monkeys had. Quietly and slowly, we unwired the cages and released the first group. What a contrast to Cripple's release! Some Wabaya animals sat motionless, glancing at the open door. Others sauntered out of the cages as if this were an ordinary place and day.

We had put the caged males at the foot of the sleeping rocks we thought were best for the troop, and had strewn the area with baboon foods. The first group slowed up to feed, as did the next group. By the time the entire troop, with the exception of the males, had been freed, I was delighted with our new procedure. The monkeys dispersed happily, unlike the frightened, tight cluster of Cripple, and were soon feeding and grunting their way up the cliffs.

That first night, Wabaya (we would have to change their name—it wasn't good public relations to introduce them to the area as the "bad guys") didn't sleep on the rocks we had chosen but went to an adjacent sleeping place with a higher vantage point. The next morning the maleless troop sunned and rested, waiting until afternoon to descend and be with the males and to examine the food we had placed around the captives' cages. Robin hopped on top of Bahati's cage, and he stood up inside to groom her bottom. Babies went to sit near their captive male friends and sought them out when larger juveniles bullied them. Although the mesh prevented a male from acting on his threat, a bluff sufficed; the juvenile would retreat and the infant would reach through the wire to groom his protector.

Late in the afternoon, rested, relaxed and satiated on the food that we had provided, the troop ascended the sleeping rocks of Ndorobo Ridge just above where the males were caged. What followed would have made Charlie Chaplin proud: although the local baboons found the impressive rock kopjes that dot the Chololo landscape suitable sleeping places, Wabaya was obviously dissatisfied. One baboon after another selected a sleeping spot, wriggled uncomfortably, shifted from side to side and then gave up and went off to search for a new resting place. The sight of so many monkeys fidgeting back and forth across the dramatic rock faces was hilarious. But the light was failing, and soon they were forced to make do.

We had another, more serious problem: How were the monkeys to get water? A dam less than a mile away was clearly visible from the sleeping rocks, so it hadn't seemed necessary to release the troop right at the water's edge, as we had done with Cripple. Also, if we had released them there, Wabaya might prefer another set of sleeping rocks closer to the dam. Since this was to be the Pumphouse Gang's release site, our challenge was to prevent Wabaya from claiming those rocks before Pumphouse arrived.

The troop had been watered before release, but by now had had nothing to drink for two days. It was hotter here than in Kekopey, and much of the food we were providing was dry. Even though the monkeys needed water badly, they made no move in the right direction. By this afternoon, they were trying to get at the water in the captive males' cages. Why didn't they go to the dam? Didn't they recognize the large, shining mass of water? Or was a mile too far for them to travel over unfamiliar terrain?

Originally I had wanted to release the males only after the rest of the troop had settled into a definite daily pattern that included both a sleeping and watering site, but perhaps the troop would be less timid if the males were with them. Which was more important: a drink or the possibility of the males leading the troop farther afield than I wanted? Again, Josiah and Hudson devised a solution: Why not try to lead the troop to the dam by laying a trail of bananas? The monkeys loved them, and so far we had been feeding them mostly with maize and cabbage. A plan was worked out, the trail laid and the troop began its banana orgy. But despite the fruit, they lost their nerve and returned to the males. Later in the day we tried again, but the troop still refused to go any closer to the dam.

We would have to release the males. Given their less conservative nature and the added protection the troop felt in their presence, perhaps they could make the trip to the dam together. The Cripple Troop experience had taught us a lesson, and this time we decided to free the males when the rest of the troop was actually around the cages. This way there would be no chance that the two segments of the troop would separate. Hudson set loose first one male, then another. Each left his cage in a relaxed way, as if he had never been captive. But Sterling, our oldest male, was a problem. For some reason, he was afraid of Hudson—a feeling that was mutual—and wouldn't let him close enough to undo the wire. I was surprised, because Sterling was a friendly old codger, a lover of babies, a finesser of females, all bark—on the rare occasions when he did bark—and no bite, because he had lost most of his teeth. I took over from Hudson, and opened the cage door. Sterling looked at me and came out. As I followed him up the rocks, I saw how delighted the troop was to have their males back. Female and infant friends rushed up to their own particular male, sat close, grunted and groomed together.

By the next day everyone, humans and baboons alike, was concerned about water. Though the freed males didn't share this concern, since they had been fed and watered amply in their cages, they did explore more widely than the rest of the troop, but not enough to solve the problem. That evening we set out large basins of water so that the thirsty baboons could have a drink. The following day Hudson decided to try the old banana trick again, and to the delight of us all the troop followed their Pied Piper to the dam and drank deeply.

I had to leave; the final, most important troop had to be captured. We divided our team, leaving some behind to keep track of Wabaya while Hudson and Josiah returned with me to tackle Pumphouse. Between us we should have enough knowledge, experience and creativity to meet whatever new challenges this troop might throw at us.

———————

Driving back to Gilgil in the bone-shattering Suzuki gave me time to mull over recent events. What exactly had we learned by moving Cripple and Wabaya? We now knew how to entice the animals to the traps, how to sedate and transport them to the holding area, how to ease the trauma of captivity, how to use food as a diversion from boredom and frustration, how to prevent injury, especially to small infants, and how to release the animals. Also, Arnold had taught us that we always needed to be one step ahead.

As I reviewed the two moves, I was suddenly struck by a frightening thought. We had managed to move all of each of the previous troops except Arnold, which was a unique accomplishment. *Why* had we been so successful? First, we knew the animals well and could prepare them properly before setting the traps. When they had showed some reluctance, we had lured them in new ways, with better bait or army garbage. Part of the success—and this was most relevant to Pumphouse—was clearly the result of having more traps than animals. It wasn't so obvious with the Cripple Troop move, but the late arrival of the Wabaya animals and their increasing reluctance to enter the traps as fewer and fewer remained unsprung suggested that if we had only as many traps as individuals we might not have been so lucky. While professional trappers set traps on several days and managed to get a few animals each day, our monkeys were different; they were smarter, wiser to the tricks of humans. I felt sure that we would get no monkeys at all if we were forced to set the traps a second time; if the baboons saw their friends and families trapped and hauled away, they wouldn't let it happen to them, too.

If we were to get the entire Pumphouse Gang at one time, we had to have a safe margin of extra traps—and that was the problem. Pumphouse was the biggest troop, with fifty-seven individuals, and we had only forty-nine traps. Granted, mothers with small babies could be counted as one individual, but there were only five such pairs. Also, some of the traps invariably misfired. We simply had to get more traps. With the help of the IPR staff, we got hold of three; this would improve but not solve the situation. We would have to try to capture some of the troop before the actual trapping began. Two days before the traps were to be set we started darting. From our past experience with

Arnold, I knew that trying to dart animals after we had trapped most of the troop would be impossible, but by doing it beforehand we hoped to be able to pick off any baboon who might be lingering on the edge of the troop, and to take the sedated animal away to the holding area before the rest of them noticed, thus causing a minimum of disruption. We didn't want to scare them and jeopardize the success of the entire operation, so we decided in advance to suspend the darting at the first signs of distress.

The Pumphouse move began early in the morning on September 25. Hudson had good shots at eight animals, but we were able to pick up only the first, fourth and last of them. The rest didn't get sleepy enough to risk handling them or injecting them with additional sedative from a hand-held syringe. The situation worsened by afternoon; the sedative had no effect on three of the animals darted, and the troop, particularly the darted animals which had not been removed, was getting restive. Later we discovered that the sedative mixed for us was of the wrong concentration; we were lucky to have captured even three individuals.

We devised another plan. Several of the adult and large adolescent males visited a storeroom adjacent to the Red House, where we kept monkey cubes and maize; it was impossible to secure it properly. We would try to take advantage of the monkeys' fondness for raiding and trap them there without the knowledge of the rest of the troop.

I had learned a great deal about the greed of male baboons, and I wasn't disappointed this time. Norman appeared first, delighted to find a whole maize cob in one of the traps. *Clank!* Down came the door, and we rushed to sedate and remove him before anyone else arrived. Just as the Suzuki was driving off, Sundance sneaked away from the troop and spotted another enticing ear of maize. *Clank!* Another victim. Oscar, Pinocchio and Wren arrived together later in the day, and were trapped simultaneously.

---

We were ready to tackle the rest of the troop; we had enough traps, a trained team and two successful captures behind us.

All went smoothly. The troop arrived early, and within fifteen minutes thirty-two of the traps had been sprung. But in that short space of time, Heckle had managed to escape from his trap, and several others had misfired. Another big male, Lou, was still free. He was as wary as Heckle, even though he had not yet been caught. Both of them were trying to get maize from the sprung traps and were cautious about entering an unsprung cage. Lou was in a quandary. He was in consort with Quanette, who had been trapped. Caged beside her were Lou's best friends—the adult female Jessie and her infant, Jordan—but Lou's mind was relieved on at least one score: Heckle was the only free male and he was busy stealing as much maize as he could without entering a trap. He was not about to challenge Lou for Quanette. So Lou ate his fill, returning frequently to the female to greet her and reassert his possession by his closeness. Finally Heckle made a mistake and sprung a trap. We rushed to secure the door with a double loop of wire so that he wouldn't elude us again. We were left with Lou.

We couldn't wait any longer; I called in the IPR team to sedate and remove the animals. There was some risk to this, because Lou would try to defend those of his friends and other troop members who gave signals of distress while being sedated. We had to keep constant watch to protect the capture team; Lou was probably all bluff, but if he decided to make good his threats, his impressive canines and great strength could do considerable damage to any human victims.

When the time came to load up the last of the animals, Lou was still free. We decided to leave Quanette and Jessie and her baby in their cages and surround them with baited traps in the hope that he might venture into one of them. In the meantime, we would feed and water the captive animals and proceed with processing the rest of Pumphouse at the holding area.

To my delight, Lou entered a trap and pulled on the bait several hours later. Unfortunately the platform was too low and he was able to execute the entire operation without having to go all the way into the cage. As a result, the trapdoor bounced off his back and he eluded us once again. We assigned staff members to watch him, and went off to the holding area. By late afternoon, he was trapped, but by the time we arrived he had escaped again. Although the door had been wired shut as soon as he had entered it, he had found a weak spot in the cage and ripped his way out. The day ended with Lou still free and the rest of us discouraged.

We had to set a time limit on Lou. The IPR team were ready to leave, and we still needed samples from the two caged females and the infant. I had been impressed and exasperated by Lou. He had discovered how to hold open the trapdoor with one hand while reaching in for the bait with the other. Systematically he sprung the traps as fast as we could set them and made off with all the bait. I'd had one shot at darting him—I'd been the only person close enough to try—and missed my chance. But watching him had given me an idea: if we strung bananas high on the inside of the roof of the strongest cage, we could force him to enter it all the way in order to reach the fruit.

Lou was interested in the bananas and tried to reach them, but when he discovered that in order to get at them he'd have to go the whole way into the trap, he backed out. I had one more idea. We could control the door by a long string, which would drop the door the moment he entered the trap. We ran the string over several cages so that he couldn't figure out the trick, even though he was watching us the whole time. It was hard to tell who was more surprised, Lou or us, when the tactic worked. The moment the door dropped, five people raced to the cage to wire it shut. Lou quickly assessed the situation and resigned himself to his fate. Then he casually reached up, pulled down a banana, peeled it carefully and ate. We let him finish before jabbing him with the syringe.

We had managed to move all three troops except for Arnold, but it had taken the sum total of our wits, experience, creativity and patience.

---

By now the biomedical processing of the baboons was honed to a fine art, and our holding procedures seemed to minimize the stress of captivity. It turned out that Pumphouse was in the worst condition of the three troops during the drought, probably due to their active exclusion from the army garbage dumps. The IPR team had rated most of the Wabaya baboons as being in good or excellent condition. Cripple Troop was in good shape, too, but Pumphouse members seldom achieved even a "good" rating, except for the males who were the Pumphouse raiders. Lou was the heaviest, weighing in at sixty-four pounds, compared with David, who never raided and weighed only forty-six pounds. Lou's coat was the thickest, and his overall appearance impressed even the IPR vet, who was accustomed to seeing overfed captive males. Male mouths were still a mess, increasingly so with age.

A few females had serious uterine infections, caused by miscarriages or births in which the cervix had never closed properly. These were the females who weighed the least for their age and showed other signs of being in poor condition. Thea's milk production was apparently affected, and this, compounded with the drought, had caused her baby, Thistle, to become dehydrated.

The IPR staff did what they could to treat the acute problems. We had agreed that such intervention was appropriate and could be important if the animals were to face successfully the stresses of their move.

Pumphouse looked worse than Cripple or Wabaya, but no worse than they had at various times over the last decade. Did this mean that they had always been in poor condition? How did their condition affect other aspects of their lives? None of the baboons we examined acted as if they were sick or in pain: not dehydrated Thistle, not the females with uterine infections, not the males with terrible abcesses. What did health mean for a baboon? More than ever, I wished there was some direct way to communicate with them.

---

As the sleeping sites and landmarks of Chololo appeared, I was delighted to spot a green flush of vegetation. There had been a brief rainfall several weeks earlier, and the land had revived; though there still was not enough food to help the cattle or most of the wildlife, there was plenty of new growth for the baboons to eat.

We lined the cages up at the dam facing the rocks and prepared for the release. Mary and I watered and fed the animals. Daylight was already beginning to fail when we evicted the IPR staff from the vicinity and released the troop, gently and slowly as with Wabaya, with a trail of food leading to the rocks—and, incidentally, to the males still held captive.

The first group made a right turn instead of going straight ahead. The second, third and fourth groups followed, and by the time all were freed it was obvious that Pumphouse was not going to go where we wanted them to. Even though it was late, they headed across the open plains and scrambled up the steepest portion of the ridge. Perhaps there *was* a pattern: all three troops had gone to the highest point they could reach before dark rather than to the better sleeping sites we had selected for them.

Now it was up to the baboons, and it seemed to me that there was a kind of poetic justice in it all. The monkeys were exerting their own preferences, demonstrating that no matter how actively I structured their world for them, they would do what *they* wanted. I stumbled back to the males in the enveloping darkness. Only a few more hurdles remained, and I had a growing confidence that all of us would make it.

# 19. Freedom

The Pumphouse Gang had been freed. What were they going to do? Would they find water and enough natural foods? How would they react to Wabaya? To the local troops?

The first day turned out to be full of delights and surprises. The freed troop headed back toward the males, but then got frightened. We fed them where they had stalled, which was close enough so that the captives heard the troop and called to them. Pumphouse raced over to their males, and the reunion chorus was unlike anything I'd ever heard: Slow grunts, fast grunts, staccato grunts, long-drawn-out male grunts, high-pitched or low-pitched, depending on the size and age of the baboon, echoed round the male cages. It was the closest I'd ever heard to baboons saying something, and what they were saying was clear: We are glad not to have lost you.

As with Wabaya, the mesh barrier presented little obstacle to social contact among friends and family. Babies went to sit close to their male friends and females wedged themselves in as best they could. Eventually the reunion chorus subsided, but a constant series of grunts continued to provide a reassuring background for resting, grooming and feeding.

Suddenly Jessie, with Jordan on her back, headed for the dam alone, quite unnoticed by the rest of the troop. Only when she returned did others pay any attention, welcoming her back with the normal grunting greeting. Jessie tried again, this time attempting to influence the rest of the troop to follow her, but without success. Twenty minutes later, Siobhan took courage and joined her aunt halfway to the dam. Still the troop hesitated. Berlioz's younger brothers came to his cage and drank from his water dish: they solved their problem the easy way, but not everyone had such a ready answer.

Nearly two hours later, Siobhan led a willing troop to the dam. They moved quickly, slowing up in the open areas to feed, but increasing their pace as the cover became thick and bushy. As I emerged from the thicket that surrounded the dam, a long row of bottoms greeted me. On Kekopey, only a few baboons could drink from a trough at one time, and natural sources of water were small. Baboons don't like to get wet and the dam was shallow at its edge, the water much muddier there than a few inches farther out. A few small peninsulas permitted both dry feet and clear water. Almost everyone found a convenient place next to a friendly neighbor and had a long drink. Still, the troop betrayed its nervousness by not dawdling. As soon as they had quenched their thirst, they headed back quickly and in single file to the males and the kopje behind them.

We had provided a significant amount of food for Pumphouse; nonetheless, it was reassuring to see that they fed on natural foods whenever they moved from place to place. Peace and contentment reigned. Satisfied baboons napped in the shade, relaxed with friends and family, groomed and played. The males were not yet free, but all else seemed well with the world.

Then Wabaya arrived. Had they heard or seen Pumphouse from their vantage point across the plains? Had they somehow noticed Pumphouse had bananas and pineapples and been unsatisfied with their own maize and cabbage? Or were they looking for someone to bully? Although Wabaya's members were fewer, they had managed to exclude or dominate Pumphouse at the army dump. Perhaps now they wanted to reaffirm their superiority.

If so, they were not disappointed. Pumphouse females and half-grown adolescents clustered tightly together and began to run away from their males, who were powerless to help them. This gave Wabaya added encouragement and they began to mob Pumphouse in earnest. A few invading adolescent males even began to chase Pumphouse females. Ordinarily I would have let the two troops settle the matter for themselves; eventually they would have to. But with the Pumphouse males incarcerated, it seemed an unfair and risky confrontation.

A shower of food fell on the rear of Wabaya, distracting their attention, as it was meant to. We slowly capitalized on a baboon first principle: Stomach Above Everything Else. We led Wabaya away from Pumphouse and the sleeping site, back toward their own rocks.

I had barely relaxed, assuming the crisis had been averted, when a local troop arrived at the kopje to settle in for the night, believing it to be uninhabited. Not only did they find a collection of caged male baboons and distraught uncaged females; they also discovered all the food that Pumphouse had deserted in its hurry to escape from Wabaya. Although most of these indigenous baboons were afraid of humans, some of the more daring males made a beeline for the food we had provided. This could be a disaster and might make it impossible to carry out our provisioning scheme for the translocated troops. There was only one thing left to do: release the Pumphouse males to give the troop a chance against the invaders and at the same time try to chase them away from the food.

The release of the males was fast and efficient. They joined the main part of Pumphouse and headed away from the cages, the food and the resident troop as fast as possible.

---

What most impressed me about the baboons during those first few days after the release was their intelligence. They were incredibly good botanists—as they needed to be, in order to support their opportunistic life-style. First they headed for the familiar: the flowering acacias and patches of star grass, both mainstays of their Gilgil diet, but in between they nibbled and sampled a variety of totally new foods. They also seemed capable of important generalizations: "Trees have fruit at the end of their branches. This is certainly a tree and there are bright round objects hanging at the ends, so they must be good to eat." Most of the time they were.

Sometimes the logic came readily but the wherewithal didn't. Pumphouse, I'm certain, had never seen a scorpion before, but they were avid consumers of invertebrates. The Chololo baboons did eat scorpions, but knew enough to pull out the stinger beforehand. When Pumphouse females and juveniles discovered one of the insects, they tried to eat it, but the scorpion would lash out with its tail anytime the nose and mouth of an interested baboon came too close. It resembled a series of tests; first one baboon would try to eat the scorpion and would finally give up in frustration; then the next in line would reenact the same routine with no greater success. After six monkeys had had their turn, the scorpion scuttled away.

I saw adults as well as juveniles consume food they had never eaten before and yet ignore tidbits that to my semi-educated eyes should have been relished. Some acacia trees had succulent growing tips to their branches. On Chololo, I was startled to see Pumphouse eating these; they had never touched them on Kekopey.

The mysteries increased. It took me nearly two hours to discover what the baboons were eating in the sandy gully leading to the dam. Nothing was visible above the ground except odd little stumps, but the area was swarming with baboons. They were eating a type of mushroom, the main part of which grew underground. This, too, I was certain they had never seen at Gilgil, let alone eaten.

Equally intriguing was the manner in which they explored the area and enlarged their home range. We were still provisioning them every morning, but by the fifth day after their release we had reduced this to a minimum, since they were finding plenty to eat on their own. We would spread six or seven pounds of monkey cubes over a wide circle, and in less than half an hour these were consumed. There was little fighting, and after eating, resting and grooming for a short period, the animals would immediately start off on a normal foraging trip, tasting whatever was around.

Each day the troop made wider circles out from the central sleeping and provisioning area, following the distribution of a certain species of acacia whose flowers represented a major part of the day's diet. They seemed so at home that it would have been difficult for an outsider to tell that these animals had not spent their entire lives here.

I had expected some kind of reaction when the troop saw an elephant or a giraffe for the first time; neither of these lived on Kekopey. But the first new item on the agenda turned out to be mating frogs. These inhabited a small temporary pool in one of the kopjes to which the baboons went to drink before starting on their day's journey. My attention was first attracted by the sound of gecking juveniles. Such gecks usually indicate mild fear, and I went to investigate. The monkeys were looking into the pool. At first I couldn't see anything, and wondered if they were reacting to their own reflections, yet I had never seen them do so this way before.

Suddenly two mating frogs emerged, piggyback. The juvenile baboons moved closer, clutching one another. When the frogs disappeared under the water's surface, the youngsters cowered, apparently more baffled than frightened, until another piggyback pair emerged nearby. Emboldened by the presence of their comrades, several juveniles ventured closer, but raced away gecking when the frogs jumped briefly out of the pool, still riding piggyback, in the direction of the baboons.

What can you expect from a bunch of juveniles? I asked myself. They had once made a fool of me in a similar fashion. Youngsters had surrounded a bush, embracing one another, looking and gecking in apparent fear. Certain that it must be a snake, I crept cautiously to get a closer look and found an old tennis shoe.

Now Berlioz and David, both serious, mature adult males, reacted exactly the same way toward the frogs. It is a rare sight to see a big male geck. When it does happen, it is usually caused by the tension of an encounter with another male who is getting the upper hand. Yet here were impressive canines exposed in a gecking gesture at two mating frogs. How embarrassing! It was was even more so when the entire troop abandoned the waterhole to these strange beasts.

---

Since the day we had released the males, native baboon troops and Wabaya had been visible daily to Pumphouse. Sometimes they even shared sleeping sites. The interactions were mostly tranquil; in fact, two of Wabaya's adolescent males actually transferred to Pumphouse. Life was really returning to normal; the frequent movement of young males between Pumphouse and Wabaya had been common at Gilgil when males were trying to make up their minds about transferring troops permanently. It now seemed that this had been broken only briefly by the translocation.

The social world of the monkeys remained almost unchanged; if anything, the bonds of family and friends were intensified and the troop became more socially cohesive. The same individuals could be found together, but now in an almost unbroken pattern. Families moved more as a unit during foraging and traveling, and rested together more consistently. The most touching of these family groups were the ones in which the mother had died, leaving behind only immature members in the matriline. These groups had always been especially solicitous of one another, but now, whenever there was a rest period, I would find them entwined in one another's arms or using each other as a pillow or backrest.

It was much hotter at Chololo than in Gilgil, and the troop rested in the shade at midday for extended periods. Rather than seeing groups of threes and fours sitting close, it was common to find clusters of nine or more baboons, all looking relaxed and at peace. Perhaps it was this heightened sociality that conveyed such a feeling of contentment to me.

I found a comfortable spot and lowered myself into a baboon position. Seen through baboon eyes, Chololo was wonderful. After the rains, the gullies, plains and kopjes would be filled with a wide variety of their foods, probably more than at Gilgil. There was plenty of water and enough sleeping sites; the locals, both baboon and otherwise, didn't seem too unfriendly and the monkeys had begun to make themselves at home. It was all too good to be true.

But was it? The short rains had come, and with them enough baboon food to allow the troops to be self-sufficient for a few months. We had stopped provisioning them about six weeks after the release, as soon as it seemed that they could find enough to eat. But the rains that brought such welcome relief from the drought were not as good on Chololo as they should have been, and by early January 1985 the lush grasslands, productive gullies and kopjes had dried up. Would the animals continue to find enough to eat on their own?

I seldom underestimate baboons, and during the translocation I had learned that they were even more adaptable and smart than I had expected. When we had originally surveyed Chololo and the surrounding area, I was disappointed that we would be so close to the Ndorobo Reserve, a tribally owned area of several hundred square miles inhabited by Samburu pastoralists and Ndorobo hunters-turned-pastoralists still living the life of their ancestors, wearing exotic garb, living in exotic settlements and having even more exotic customs. The place was a wasteland, an arid region of bare ground between thorny bushes and shrubs, and I wondered what the goats and cattle found to eat. There was no constant supply of water, the heat seemed more intense, and though the landscape had a stark beauty, I couldn't help thinking it was an area that produced more rock than anything else.

But it was the rock that formed part of the explanation for the rapid change in vegetation over such a short distance. Chololo had wonderful grasslands and a variety of inviting trees and bushes. Only several hundred yards from the boundary, the Reserve produced stunted plants, little grass and impressive thorns. The extraordinary difference was due to the depth of the soil and its rockiness; the vegetation in the Reserve was specially suited to the sparse soil. When we had traveled through it earlier, I had mentally written it off as baboon country. I was certain the animals would look southward to a nearby ranch much like Chololo, abundant in grassland and fruiting trees, but strictly off limits to the baboons and to us. Its owners did not like baboons or baboon watchers.

My heart sank when, fifteen weeks after the translocation, Josiah reported back to me in Nairobi that the troops were in the Reserve. The research assistants were all worried, and on my next trip to Chololo I approached the monkeys with a heavy heart—only to discover that they had found a dry-season gold mine. The Reserve was an arid-adapted system; while the plants didn't look impressive, they were well suited to surviving the dry conditions, storing important nutrients and water in their stems and leaves.

I followed the animals, spellbound. Watching Robin's family closely, I began to understand how baboons constructed a balanced diet from such a variety of new and unusual foods. A major part of it was succulents, particularly Sansevieria, which they had never encountered before, and my old friend the prickly-pear cactus—Opuntia. Several species of acacias abounded in the Reserve. Although the baboons didn't seem to like or eat the seeds while they were on the trees, as they had with the species that grew at Gilgil, now they searched for and ate the dried ones that had fallen to the ground or that they discovered in the goat droppings raked out of the Ndorobo settlements.

Some dried grass found here and there, dried berries and a variety of leaves from different bushes rounded out the menu on the first day of my return. Both the Opuntia and Sansevieria required a great deal of strength to harvest. The Opuntia pads had to be knocked off the stand by sheer leverage and with careful maneuvering to avoid thorns. This meant that only the larger baboons could get them. Sansevieria had to be pulled from the ground, since the baboons ate only the soft, fleshy base, not the hard part that sticks up above ground. I was curious about the Sansevieria. First I pulled at it with one hand, then with both, until finally, with all my effort, I managed to yank up one of the leaves. Although baboon youngsters are incredibly strong by human standards, it was obvious that the smaller ones were as handicapped in trying to get at the Sansevieria as they were with the Opuntia. Fortunately some of the younger leaves were easier to obtain, but in general the youngsters followed a very different feeding plan from that of the adults.

Robin, now the matriarch of a Wabaya family, marched from stand to stand of Sansevieria, selecting leaf after leaf based on mysterious criteria. She wandered here and there, finding dried grass, some roots, bulbs and dried berries, and ended up at a stand of Opuntia, where she occupied herself for several hours, demolishing a significant portion of it. Riva, her juvenile daughter who was about the age that Robin was when I first met her in Pumphouse (she looked remarkably like the young Robin who had thrilled me by trying to groom me), followed behind, pulling up small Sansevieria wherever she could. Then she went to Robin's discards, apparently finding something more to eat there. Following in Robin's footsteps, eating grasses, berries and leaves, she finally arrived at the same Opuntia stand where her mother was having an enthusiastic feeding spree.

Mother and daughter fed comfortably side by side, Riva once again managing well on what her mother had left behind. Meanwhile, little Rima, just weaned and still dependent on her mother for protection, nourishment and an occasional free ride, meandered through the open area picking up the same foods as her mother and older sister before joining them at the stand of Opuntia.

The baboons were finding plenty to eat, and obviously they were getting enough water from the Opuntia and Sansevieria. Their new diet must have been rich in what they needed, I decided, for back at Gilgil they had fed from first light to dusk in the dry season, with barely enough time to rest. Here on the Reserve, they rested for long periods during the day, were lazy about leaving their sleeping sites in the morning and retired early in the afternoon.

At the end of the day I was elated. The baboons knew what they were doing; their chances for survival were even better than before. Serendipity had intervened; they could not have been released in a better place. Chololo was a wonderful rainy season home, and the adjacent Reserve was a treasure trove in the dry season. It had taken them some time to find exactly the right places. During the first weeks of the dry season, the baboons had covered a thirty-square-mile area, searching and exploring. But they knew what they were looking for, and they had found it. Hudson, who was with them at the time, felt that they had taken some of their cues from the native baboons. Some days they would share sleeping sites, and when the indigenous group moved off in a new direction, Pumphouse would follow, explore and then decide what to do. Just how they developed a taste for new foods such as Sansevieria was difficult to determine, but in Wabaya Troop the males who transferred from native troops could have played an important role, eating what for them were traditional foods and what for the translocated monkeys was a new diet.

Through exploration and observation, their diet was constantly expanding. Jessel, one of the new males in Wabaya who had come from the Chololo population, dug for salt in a dry riverbed. Zilla followed his lead, and several other troop members were soon eating from the same source. Although I continued to make plans for future provisioning in case the long rains failed, I felt with each successive day that the baboons would survive, making a place for themselves in their new home.

Pumphouse resumed what seemed in all its aspects a normal life, not the life of desperation before the move, but the life of an earlier period. There had been no deaths in the sixteen weeks following the translocation. This was in strong contrast to the mortality rate on Kekopey, when the baboons were forced to visit the army dumps at the height of the drought and when thirteen animals disappeared or were killed in as many weeks. Injuries, all too common when the troop relied on army food, were now almost nonexistent. Although Zilla had given birth to a stillborn infant and Beatrice had had a late miscarriage, both clearly the result of the trauma of capture and transport, six new babies had been added to the two troops since the move, babies conceived at Kekopey.

But a normal life doesn't simply mean new infants and healthy animals; death and decline are part of the natural life cycle. In a population in equilibrium, as many animals die as are born, and not every individual can be healthy. Charity disappeared the day I arrived, leaving Constance with one less daughter and David without his most devoted infant friend. What happened to her is not known. She was healthy one evening and gone the next morning. The troop was particularly nervous at its sleeping site, and the local people had reported seeing a large snake on that kopje. Certainly an accident had occurred, but that was part of baboon life.

Several adults were in poor condition. My old friend Sterling seemed to be on his last legs, lagging behind the troop when they left in the morning and sometimes traveling alone. Each night we were convinced that he had died and were delighted and relieved to find him each morning, feasting on acacia seeds hidden in a pile of goat dung.

Constance and Vicki were now the oldest Pumphouse females; Vicki was at least thirty-four years old. Both looked disheveled from the side and rear, but still appeared well groomed from the front. When Peggy died, the condition of her teeth indicated that she could have continued to chew her food adequately for only a few more years; after that she would have lost weight and succumbed to disease. Vicki was rapidly approaching the same state.

Although the death of any of the baboons was a sad event, I had learned to accept it as part of the natural cycle of life. The impending deaths and Charity's disappearance were easy to take compared with the heavy mortality rate among the troop during the crises at Kekopey over the past five years.

I left the baboons again in March 1985. This time I had the feeling that all was right with the world once more. I had given them their best chance; they had seized it and made the most of it. The future would not be without problems, their lives would not be without snags, but they were no longer the victims of overwhelming circumstances. I was convinced that they would now survive, and that we would be able to study them and gain new and exciting insights, both about them and about ourselves. I felt that I had finally fulfilled my part of our relationship. Unwittingly, the baboons had given me much over the years, personally and scientifically. Now I had been able to reciprocate by giving them a chance for a better future.

I glanced back at David and Constance and Thea and her family as I hurried to reach the car before dark. I was overwhelmed by a feeling of optimism and, for the first time in the decade since the troubles began, with a sense of freedom for both the baboons and for me. If our lives intertwined hereafter, it would be out of choice rather than from necessity. Now, if they wanted to, it would be possible for my children to watch the Pumphouse Gang too.

APPENDICES

# Appendix I

## Communication

Baboons communicate by means of a series of gestures, postures, facial expressions and sounds. Unlike human language, baboon sounds are not symbols referring to objects. For the most part, the sounds and other communications that baboons exchange express their emotions: *I feel angry, I feel happy, I feel content, I am confused, I am ambivalent, I am not aggressive* and so on. Some sounds may also contain a small amount of other information, as has been demonstrated in vervet monkey alarm calls and grunts. Some baboon sounds—grunts, for instance—may give information about the individual at whom they are directed: a dominant female, a stranger or a baby.

Like all monkey and ape communication, baboon communication is both graded and repetitive. Signals combine to form a series that reflects the intensity of the animal's emotion: *I'm a little angry, I'm moderately angry, I'm very angry.* Sounds, facial expressions, body postures and gestures are used simultaneously to emphasize a point or to add degrees of intensity to it.

It is difficult to imagine how simple statements of emotion alone can communicate the complexity of baboon interactions, but these signals are not used by themselves; they are produced in a particular social and environmental context, and it is this context that provides the actual meaning—meaning that can be as varied as the situation.

All animals need to communicate with one another. Solitary creatures who reproduce sexually need only to be able to communicate about mating. Creatures who care for their offspring need to be able to exchange information about themselves and their children. Animals who live in social groups, temporarily or permanently, need to be able to communicate on a much more sophisticated level. This is probably how communication evolved. As a result, signals were borrowed from one context—sexual behavior, for example—and used in another—parenting or social communication—with a shift in meaning to extend the dialogue. Primate researchers at the turn of the century failed to understand this sequence and assumed each signal had only one possible meaning. Their interpretations included many ideas that today we recognize as erroneous.

These drawings illustrate the visual range of baboon signals (which, of course, are also accompanied by a wide variety of equally important sounds). Each sequence portrays the same signal in different contexts. Comparing contexts helps to convey the ways in which what is communicated can be both similar and different.

# Presents

A sexually receptive female presents to a male, exposing her bottom for the male to inspect (Fig.1). This is a communication about sexual intentions. If the time is right, a sexual present can be a prelude to a copulation or even to the start of a consort relationship.

Presents also occur outside sexual contexts. In Fig. 2, a female presents to a male as a form of simple greeting. A male's higher social status requires that some statement be made if a female enters the large personal space around him that he considers his own.

Fig. 3 is a variation of a female greeting to a male, but in this case the female is actually frightened of the male. Her greeting present is from a distance and her tail is held much higher, even bent toward her head—always a clear sign of fear.

Presents can be invitations to approach, as in Fig. 4, where an adult female presents to an infant, inviting it to come close. Or a present can be a greeting asking permission to approach, as in Fig. 5, where a juvenile female presents to a mother as a prelude to trying to fondle her infant.

*Figure 1*

*Figure 2*

*Figure 3*

*Figure 4*

*Figure 5*

# Variations of Presenting

Males sometimes greet each other by actual presents of the female type—tail up, bottom directed to the other animal—but more frequently male greetings are mutual and incomplete as far as the tail-up position is concerned. In Fig. 6, two males are side by side, tails only partly different in orientation; notice, however, that each male is facing and looking at the other male's bottom, as in a normal present. Other signals are also being exchanged: the males are grunting at each other, narrowing their eyes and smacking their lips together. The signals convey friendly intentions but also some nervousness.

Presents can be part of more complicated communication. In Fig. 7, a juvenile presents in a greeting gesture to an adult male while at the same time he is acting aggressively toward another, larger juvenile in the distance. This combination is often called a "protected threat," since the individual is counting on the help or at least the protection of the adult male, giving it the courage to threaten a feared or more dominant opponent.

Fig. 8 illustrates a very different type of present—a present for grooming. Here a male is presenting the part of his body he wants to be groomed by the approaching female. This is both an invitation to approach and a grooming request.

*Figure 6*

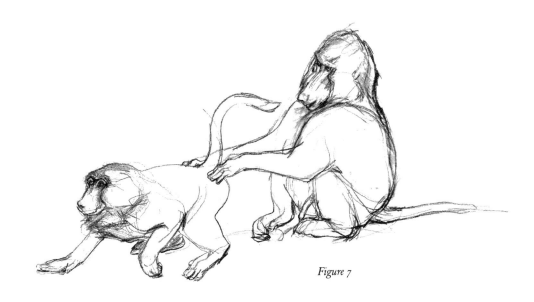

*Figure 7*

*Figure 8*

## Embraces

Embraces are one type of greeting. Animals who are not related to one another or who do not associate frequently find embraces a useful way in which to convey friendly intentions. Embraces probably originated from handling infants, as shown in Fig. 9. The female—not the mother—fondles the young baby gently, bringing it close to her chest and face. Frontal embraces are an appropriate greeting between a female and an older infant (Fig. 10) and between an adult male and an infant (Fig. 11). Fig. 12 shows two adult females embracing. Adult embraces are more often sideways than frontal, although frontal embraces occur when individuals are also trying to reassure each other after some particularly upsetting incident. This type of behavior was noted during the translocation of Cripple Troop (see page 239). Among adults, the frontal embrace appears almost an infantile regression.

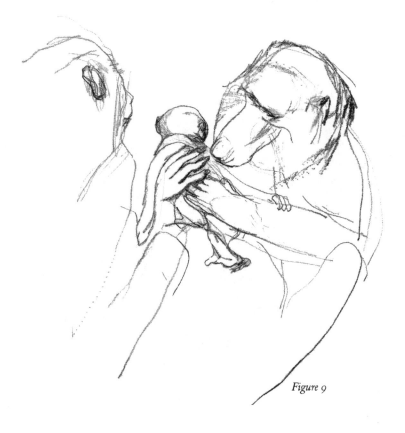

*Figure 9*

*Figure 10*

*Figure 11*

*Figure 12*

# Aggression

Communicating aggressive intention is the most clearly graded of baboon signals. At its lowest intensity, a simple lifting of the eyebrows reveals white eyelids, signaling that the sender is aware of what is going on and is not happy about it (Fig. 13). In Fig. 14, an adult male is intent on letting those around him know how he feels; he is much more upset. He has added an open-mouth threat to the eyelid signal, revealing impressive canines. His hair is also beginning to stand on end, making him seem even more formidable. As the threat heightens in intensity, more sounds and ground slapping will be added. Fig. 15 shows a juvenile male equally aroused; the white eyelids are visible, the hair is standing on end and the mouth is wide open to reveal absent canines. Baboon signals are ritualized. While the open-mouth threat was originally probably a way to display the force behind the bluff, its meaning is now less dependent upon actual flashing of impressive canines.

Aggressive signals have their counterpart in submissive ones. Sometimes presents and other forms of greeting are used to appease or to stop threats. At other times, as in Fig. 16, screams, fear-faces and crouching low demonstrate how upset an animal is. The female at the left is lunging toward a cluster of youngsters, giving an eyelid threat and making aggressive sounds; she is about to slap at them. The juvenile directly opposite her is screaming with an open mouth and lurching backward. The infant in the background is watching fearfully, poised to move off and making a fear-face, its mouth pulled wide in what looks like a grin but which means the opposite. Accompanying the fear-face is a sound called a "geck" or "gecker," a series of short, staccato grunts related to fear. The infant in the foreground is crouching down and is about to try to hide behind a small rock.

*Figure 13*

*Figure 14*

*Figure 15*

*Figure 16*

# Play

Play is accompanied by a special signal—the play-face. Many other aspects of behavior are different in play—for example, the gait is often looping and inefficient—but it is the play-face that conveys the important message: *This is not serious, even if it may seem so.* All behavior can become part of play, including what would otherwise be construed as aggression. In Fig. 17, two black infants are just beginning to experiment with social play and have not yet mastered the signals. The six-month-old infant in Fig. 18 is playing with a much larger individual and exhibits a partial play-face. The two juveniles in Fig. 19 both show a full play-face, but are also trying to bite each other on the mouth.

*Figure 17*

*Figure 18*

*Figure 19*

## Ambivalence

Because emotional reactions are often complicated and confused, communication can show a great deal of ambivalence, evidenced either by alternating between types of signals—males often first exchange friendly greetings, then threats, then friendly greetings again—or by sending conflicting signals simultaneously, as shown in Fig. 20. Here the female on the right has approached (communicating friendly intentions) the higher-ranking female and infant on the left, and is torn between attraction to the baby and fear of the mother. The fear-face is the same as that displayed by the juvenile in Fig. 16, and appears similar to the open-mouth threat shown in Fig. 15, but in a true fear-face the mouth is not opened as wide, the lips are more curled and the eyes are not closed in an eyelid threat.

*Figure 20*

Baboons use nonverbal communication. Humans share this nonverbal system as well, although the signals are not always identical. Even when we talk, we are also communicating through posture, gestures and facial expressions; as with baboons, our nonverbal system conveys how we feel. Because we have two systems of communication, we can say one thing and mean another. The truth—whether we really like someone, whether we are really upset or not, whether we believe what we are saying—is most often found in the nonverbal messages.

# Appendix II

# Table I–Peggy: A Baboon Family *

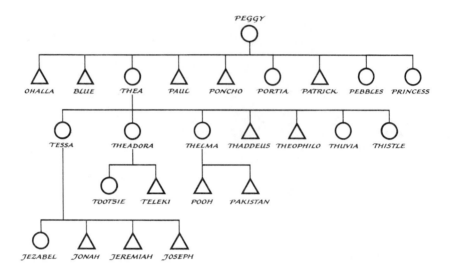

*A typical baboon family spans several generations. This table represents Peggy's family as I documented it between 1972 and 1986. In 1972, I was able to determine which were Peggy's older children—Ohalla, Blue and Thea—by their behavior patterns, but the remainder of those listed are births that were observed firsthand. Thistle, Pakistan, Jeremiah and Joseph are grandchildren and great-grandchildren born after Peggy's death. When the troop split in 1981, Tessa joined Wabaya. The rest of Peggy's family has remained with Pumphouse.

# Table 2–Peggy: Fifteen Minutes*

*Important information about baboon social behavior is obtained by following individual animals at specific times each day for a predetermined period. The data recorded here include Peggy's basic activities—feeding, resting and traveling—as well as a detailed description of the social interactions in which she was involved during a fifteen-minute time span. To place Peggy more accurately in the context of the rest of the troop, information about what other baboons were doing and the names of the baboons in her vicinity were also noted.

# Table 3—A Field Worker's Daily Notes*

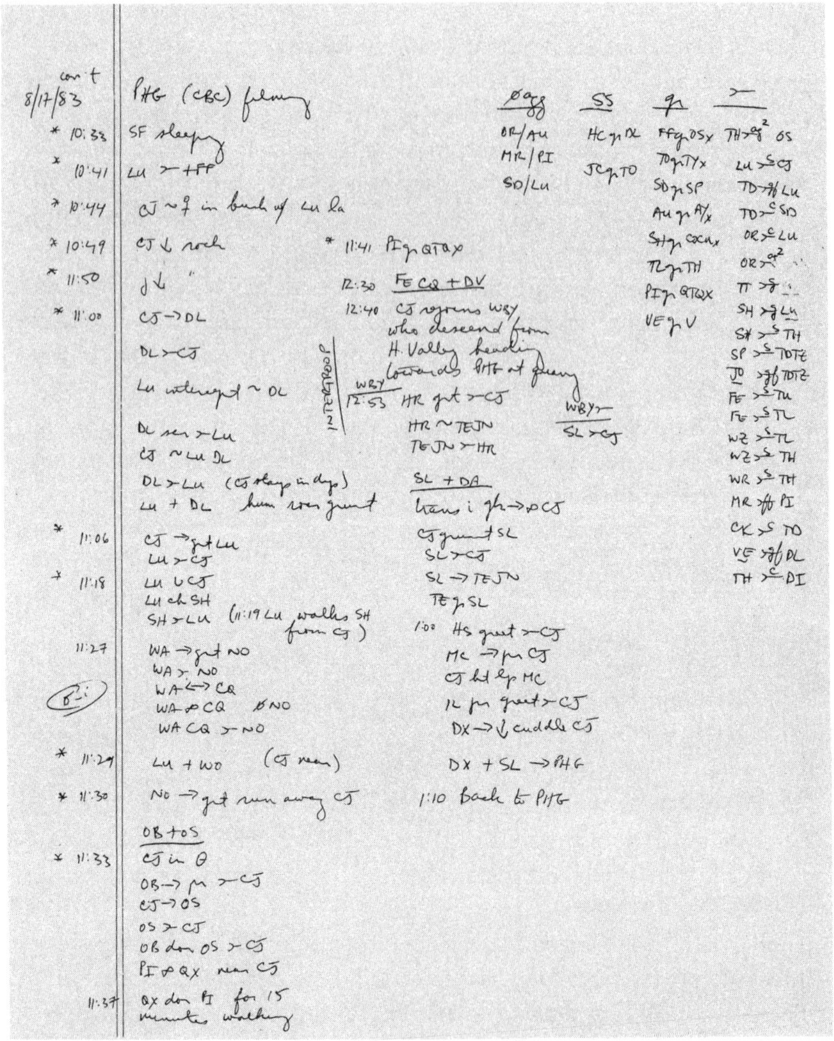

*Notes such as these, written in a shorthand I designed to help record quickly and accurately the many simultaneous, complicated events occurring in a baboon troop, give the life story of the troop a day at a time. Crucial information on aggression, grooming, dominance interactions, special encounters between friends, use of infants and females as agonistic buffers, sexual consorts and copulations is recorded on the spot, no matter which individual is involved.

# Bibliography

TECHNICAL

Darwin, Charles. *On the Origin of Species.* London: John Murray, 1859.
———. *The Descent of Man.* London: John Murray, 1871.
DeVore, Irven. *Primate Behavior: Field Studies of Monkeys and Apes.* New York: Holt, Rinehart and Winston, 1965.
———. "Male Dominance and Mating Behavior in Baboons," in F. Beach, ed., *Sex and Behavior,* New York: Wiley, 1965.
———, and Washburn, S. "Baboon Ecology and Human Evolution," in F. C. Howell and F. Bourlière, eds., *African Ecology and Human Evolution,* Chicago: Aldine, 1963.
de Waal, Frans. *Chimpanzee Politics.* London: Jonathan Cape, 1982.
Fedigan, Linda. *Primate Paradigms: Sex Roles and Social Bonds.* Montreal: Eden Press, 1982.
Griffin, Donald R. *The Question of Animal Awareness.* 2nd. ed., New York: Rockefeller University Press, 1981.

Hall, K.R.L., and DeVore, Irven. "Baboon Social Behavior," in I. DeVore, ed., *Primate Behavior*, New York: Holt, Rinehart and Winston, 1965.

Jay, Phyllis. *Primates: Studies in Adaptation and Variability*. New York: Holt, Rinehart and Winston, 1968.

Jolly, Alison. *The Evolution of Primate Behavior*. New York: Macmillan, 1985.

Kuhn, Thomas. *The Structure of Scientific Revolutions*. 2nd. ed., Chicago: University of Chicago Press, 1970.

Latour, Bruno, and Strum, Shirley C. "Human Social Origins: Please Tell Us Another Story," *Journal of Social and Biological Structures,* 9: 167–187, 1986.

———, and Woolger, Steve. *Laboratory Life: The Social Construction of Scientific Facts.* Los Angeles: Sage, 1979. (New edition, Princeton University Press, 1986.)

Lee, Richard, and DeVore, Irven. *Man the Hunter.* Chicago: Aldine, 1968.

Rowell, Thelma. "Forest-living Baboons in Uganda," *Journal of Zoology,* 149:-344–364, 1966.

Simpson, George Gaylord. *The Meaning of Evolution.* New Haven: Yale University Press, 1963.

Strum, Shirley C. "Primate Predation: Interim Report on the Development of a Tradition in a Troop of Olive Baboons," *Science*, 187:755–757, 1975.

———. "Baboons," *Wildlife News,* Kenya 1:4–10, 1976.

———. "Predatory Behavior of Olive Baboons *(Papio anubis)* at Gilgil, Kenya," Dissertation, University of California, Berkeley, 1976.

———. "Primate Predation and Bioenergetics: A Reply," *Science* 191:314–317, 1976.

———. "Dominance Hierarchy and Social Organization: Strong or Weak Inference?" Paper for Wenner-Gren Conference, *Baboon Field Research: Myths and Models,* June, 1978.

———. "Processes and Products of Change: Baboon Predatory Behavior at Gilgil, Kenya," in *Omnivorous Primates*, G. Teleki and R. Harding, eds. New York: Columbia University Press, 255–302, 1981.

———. "Agonistic Dominance in Male Baboons: An Alternative View," *International Journal of Primatology,* 3:175–202, 1982

———. "Why Males Use Infants," in D. Taub, ed., *Primate Paternalism,* New York: Van Nostrand Reinhold, 1983.

———. "Use of Females by Male Olive Baboons *(Papio anubis),"* *American Journal of Primatology,* 5:93–109, 1983.

———. "Baboon Cues for Eating Meat," *Journal of Human Evolution,* 12:327–336, 1983.

———. "Baboon Models and Muddles," in *The Evolution of Human Behavior: Primate Models.* W. Kinzey, ed., New York: SUNY Press, 1986.

———. "A Role for Long-term Primate Field Research in Source Countries," *Proceedings of The International Congress of Primatology,* Vol. 3, J. Else and P. Lee, eds., Cambridge: Cambridge University Press, 1986.

———. "Activist Conservation—the Human Factor in Primate Conservation in Source Countries," in *Proceedings of the International Congress of Primatology,* Vol. 3, J. Else and P. Lee, eds., Cambridge: Cambridge University Press, 1986.

———. "Translocation of Primates," in *Primates: The Road to Self-Sustaining Populations,* K. Benirschke, ed., New York: Springer Verlag, 1986.

———, and Western, J. D. "Variations in Fecundity with Age and Environment in a Baboon Population," *American Journal of Primatology,* 3:61–76, 1982.

Teleki, Geza. *The Predatory Behavior of Wild Chimpanzees.* Lewisburg, Maine: Bucknell University Press, 1973.

Thibodeau, Francis R., and Field, Hermann H. *Sustaining Tomorrow.* Hanover, N.H.: University Press of New England, 1984.

Washburn, Sherwood, and DeVore, Irven. "The Social Behavior of Baboons and Early Man," in S. Washburn, ed., *Social Life of Early Man.* New York: Wenner-Gren Foundation, 1963.

——— and Hamburg, David. "Aggressive Behavior in Old World Monkeys and Apes," in I. DeVore, ed., *Primate Behavior.* New York: Holt, Rinehart and Winston, 1965.

Western, J. D., and Strum, Shirley C. "Sex, Kinship and the Evolution of Social Manipulation," *Ethology and Sociobiology,* 4:19–28, 1983.

Wilson, E. O. *Sociobiology: The New Synthesis.* Cambridge: Belknap Press, 1975.

Zuckerman, Solly. *The Social Life of Monkeys and Apes.* London: Routledge & Kegan Paul, 1st. ed., 1932; 2nd. ed., 1981.


GENERAL

Ardrey, Robert. *African Genesis.* New York: Dell Publishing Co., 1961.

———. *The Hunting Hypothesis.* New York: Atheneum, 1961.

Dawkins, Richard. *The Selfish Gene.* New York: Oxford University Press, 1976.

Fossey, Dian. *Gorillas in the Mist.* Boston: Houghton Mifflin, 1983.

Goodall, Jane. *In the Shadow of Man.* Boston: Houghton Mifflin, 1971.

Gould, Stephen Jay. *Ever Since Darwin.* New York: W.W. Norton, 1977.

Leakey, Richard, and Lewin, Roger. *Origins.* London: Macdonald and Janes, 1977.

Lorenz, Konrad. *On Aggression.* New York: Bantam Books, 1971.

Marais, Eugene. *The Soul of the Ape.* First published London: Penguin Books, 1979, although book was originally written in the 1920s.

Moorehead, Alan. *Darwin and the Beagle.* London: Hamish Hamilton, 1969.

Myers, Norman. *The Sinking Ark.* New York: Pergamon Press, 1979.

Nash, Roderick. *Wilderness and the American Mind.* New Haven: Yale University Press, rev. ed., 1973.

Strum, Shirley C. "Life with the Pumphouse Gang: New Insights into Baboon Behavior," *National Geographic*, 147:672–691, 1975.

————. "Baboons Today," *Kenya Past and Present*, Vol. 12:21–27, 1980.

————. "Baboon Behavior," *Swara*, 4:24–27, 1981.

————. "Baboons: Social Strategists Par Excellence," *Wildlife News*, 16:2–6, 1981.

————. "The Pumphouse Gang and the Great Crop Raids," *Animal Kingdom Magazine*, 87:36–43, 1984.

————. "Baboons May Be Smarter than People," *Animal Kingdom Magazine*, 88:12–25, 1985.

Tiger, Lionel. *Men in Groups*. New York: Random House, 1969.

Tinbergen, Niko. *Curious Naturalists*. Garden City, N.Y.: Anchor Books, 1958.

Washburn, Sherwood, and Moore, Ruth. *Ape into Man*. Boston: Little, Brown, 1974.

# Index

Sterling, 119, 245, 259–60
Strand (ranch manager), 207–8
Strider, 48, 109, 121
*Structure of Scientific Revolution, The*
(Kuhn), 163–64
submissive signals, 272
Sumner, 21, 33–34, 65–66, 77–78, 121,
122, 129, 130, 131–32, 133, 203
Peggy's friendship with, 50–51, 52
Survival Anglia Television, 174, 200,
230

tarsiers, anatomy of, 71–72
taste-aversion conditioning, 180–83
teeth, 234
canine, 34, 142–43
Tessa, 41, 43, 52, 66, 138
sexual competence learned by, 47
Thea, 39, 40–41, 42–43, 52, 66, 87, 139,
249, 260
Peggy subjugated by, 137–38, 143
Theadora, 41, 43, 46, 52, 138, 139
Thelma, 41, 43, 44–45, 46, 52, 134,
138
Thomson's gazelles, as prey for
baboons, 65–67, 129–32
threatening gestures, 26, 34
thumbs, opposable, 70
Tim, 28, 29, 31
Tina, 112, 113, 134
Tinbergen, Niko, xiii
Tisot, Joseph, 232, 233
Toby, 133–34, 155
tools, 73, 74, 143, 150
translocation, 205–60
airplane reconnaissance in, 208–11,
220
baboons' adaptability and, 256–59
behavioral changes and, 222
biomedical processing in, 233–34,
249–50
of California deer, 218–19
captivity period in, 234–35, 243
capture and sedation operations in,
230–34, 242–43, 246–49
Chololo site in, 221–22, 224, 225–27,
243–45, 250–53

Colcheccio site in, 214–17, 219–21,
226, 235–41
of Cripple Troop, 229–41, 242, 243,
245, 246, 249
feeding in, 234–35, 236, 240, 243, 246,
251, 252, 253–54, 256–58
indigenous baboons and, 238, 259
intertroop encounters after, 238,
252–53, 255
Mbale site in, 223–25, 227
to national parks, 205, 223
new wildlife encountered in, 254–55
of Pumphouse Gang, 205–30, 244,
245–60
releases in, 236–41, 243–45, 250–53
site sought for, 208–27
sleeping sites and, 220–21, 244
social patterns unchanged in, 255–56
Strum's meetings with landowners in,
211–17, 223–27
transportation to new site in, 235–36
to tribal lands, 205, 223, 256–57
of Wabaya, 230–31, 241, 242–45, 246,
249, 259
watering sites and, 244–45, 252
white landowners' attitudes and,
205–8
traps, 230–31, 246
tribal lands, 205, 223, 256–57
troops:
daily routine of, 19, 27, 55, 179–80
emigration from, 93, 94–104, 107–8,
136–37, 151–52
females as stable core of, 45, 79,
137–39, 157
interactions between, 92–93, 94–95,
99–100, 104, 110, 238, 252–53,
255
"leaders" of, 80
males as transient members of, 77, 79,
157
movements of, 76, 80, 172–73
multi-male, 74–75
newborn infants and, 42–44
newcomers in, *see* newcomers
residency status in, 117–19, 121–23,
126, 136

splitting of, 177–78
ties between members of, 51, 52
*see also* Cripple Troop; Eburru Cliffs;
    Pumphouse Gang; Wabaya
Turner, Philip, 207

variation within species, 73
vervets, 11, 72
Vicki, 112, 113, 121, 260
Virgil, 98–99, 102, 103
vision, 70

Wabaya:
    at army camp, 203–4, 205, 242
    crops raided by, 178, 179, 183–
       84
    formation of, 177–78
    interactions between Pumphouse
       Gang and, after translocation,
       252–53, 255
    translocation of, 205, 230–31, 241,
       242–45, 246, 249, 259
Washburn, Sherwood, 14, 19, 51, 72,
    162, 164, 199, 201

aggression-language relationship as
    viewed by, 145–46, 147, 148
baboons studied by, 73, 75–76
primate evolution as explained by,
    69–71
Strum's classes with, 9–10
Strum's research proposals and,
    11–13, 77
watering, 27, 244–45, 252
weaning, 234–35, 237, 240
Wenner-Gren Foundation, 160
Western, David (Jonah), 195–96
    crop raiding and, 184–85, 186–87,
       189, 192–93, 194
    Strum's marriage to, 227–28
    in translocations, 208–11, 212, 221,
       223, 224, 225, 227
Wilberforce, Samuel, Lord, 71
Wildlife Clubs of Kenya, 187
Williams, Lynda, 14
Williams, Matt, 14, 15, 18, 21, 22, 23,
    24, 27, 60, 169
Woolcraft Project, 200
Wound, 96, 100